Wormhole *in vitro*
Big Bang model, Cronus Hyper-Capacitor and Teleporter

Author:
Dr. *Antonio Silvestro*, self-employed heavenly based.
Department of *'Mathematic, Physics and Natural Sciences'*, Ba *'Biological Sciences'* University Federico II of Naples, Naples (NA), 80100, Italy.
Department of *'Agriculture'*, MSc *'Plant, Food Science and Environmental Biotechnology'*, University Federico II of Naples, Portici (NA), 80055, Italy.

Abstract:
The author is honoured to have the opportunity to propose a cutting-edge 'Wormhole *in vitro*' in which state-matter would exchange through the Minkowski spacetime generating exceptional potential suitable for triggering the Cosmic Wave background (CWB) that have been taking part in the continuous cycle of birth, death and rebirth characterizing the Nirvana. Among its applications a realistic perspective about the abiogenesis Ab Initio Molecular Dynamic (AIMD) of the Solar System (SS), the spontaneous generation and storage of power catching sunlight from the future to enlighten the past in the bouncing present where a SuperNova (SN) found her Black Hole, the once happened in the Triassic – Jurassic (Tr-Jr) transition encrypted on the glyphs of the Aztec Sun Stone Almanac named in honour of the jaguar *Tezcatlipoca*, and a suitable theoretical treasure for the design of a human Teleporter. Nevertheless, here and now, the spacetime fringe has been instantaneously passed led to the creation of a device able to transduce the genome of unicellular organisms via the quanta. Choosing the primeval ocean like the descendant of the Last Universal Common Ancestor (LUCA) most close to it and still alive in the current Holocene, the phototrophic *Cyanobacteria*, has been highlighted that the AIMD begun before the Big Bang in a White Hole related to the Black Hole of the Solar System Supernova from which the planets, among which the Earth where we sentient being all living on for the moment (Mars displacement coming soon - 2025) like the heterotrophic fish spoilage *Proteobacteria Alteromonadales Shewanellaceae*, shedding light on a potential genesis of the water molecule (H_2O) behind the spatial mirror dated 13.8 GYA.

Keywords: battery, capacitor, solar panels, nuclear power, SuperNova (SN), *Homo atm*, Big Bang, patent, prototyping, hacking, biotechnology, astronomy, astrology, cosmology, astronomy, astrobiology, esoterism, human bioenergy, nuclear energy, biomagnetism, physics, engineering, electronics, electro-physiology, psycho-physics, singularity, Black Hole, Big Bang, wormhole, LUCA, Yoga, Kundalini, teleportation, Microbial Fuel Cells (MFC), Microbial Electrolytic Cells (MEC), transplantation, genesis, co-evolution, quantum mechanics, resonance, M-theory, Electro Magnetic Sequencing (EMS).

Correspondence for Copyright © permissions requirement to:
Dr. Antonio Silvestro born Friday 15th May 1992 at 20:00 under the Taurus sign ♉ ascendant Scorpio ♏ according the Greek specular to Monkey 猴 hóu rising dog (*Canis lupis*) Canis Major 狗 gǒu in Chinese and chestnut (*Castanea sativa*) in Druid, flower (*xochiti*) in Aztec astrology, North knot Capricorn ♑ goat (*Capra hircus*) 羊 yang and South knot Cancer ♋ rat (*Ractus norvegiensis*) 鼠 shǔ, resident in n°100 Nazario Sauro St., 80026, Casoria (NA) (Italia), number phone: +39 3382634244, emails: dr.antoniosilvestro@gmail.com, tonysilverxxx@gmail.com and antonio.silvestro5@studenti.unina.it.

'Wormhole *in vitro*' © Copyright Antonio Silvestro, 2020

Index

- Batteries vs capacitors...2
- Localizing elementary particles and their shah in the Big Bang wormhole ....…..8
- Nuclear destruction vs creation: H-Bomb vs Tokomak………………………………….19
- Cosmic Wave Background (CWB)………………………………………………....…23
- SuperNovas (SN), Milky Way (MW) and Dark matter………………………...……...28
- Artificial wormhole ……………………………………………………………………….30
- Helios irradiating……………………………………………………………………….55
- Dye Sensitive Solar Cell (DSSC)……………………………………………………....59
- PhotoVoltaics (PVs) panels…………………………………………………………….63
- Autotrophic inoculum for White Hole anode (+)……………………………………….65
- Heterotrophic inoculum for Black Hole catode (-)……………………………...……...74
- Teleportation milestones………………………………………………………………..80
- Microbial Teleporter (MT) design and architecture……….....................................84

Batteries vs capacitors

Alessandro Giuseppe Antonio Anastasio Volta is the inventor of the first **disposable primary cell battery** transducer of chemical into electrical energy, the so-called voltaic pile done of alternated layers of oxidizing anode zinc (Zn -> Zn^+ + e^-, V_{Zn} = - 110 V) and reducing cathode copper (Cu -> Cu^{2+} + 2 e^-, V_{Cu} = - 20 V) n-pairs separated by cardboard dipped into an electrolyte solution (e.g. brine solution done of sodium chloride NaCl dissolved in water H_2O) reducing friction between the metallic sheets, reminding the thylakoids of the chloroplast of the vegetal photosynthetic biological cells, otherwise, magnetic field inductors and protein secondary structures α-helixes characterized by N-turns corresponding to the number of electrodes pairs, for which the invisible frequency, pitch and energy algorithm may be utilized for designing physical innovative batteries. Certainly, you may not forget the Kundalini life triggering energy done of the same ascensional or discretional cylindrical helical path, from the past to the future, the white and black hole, and vice versa as Albert Einstein would, perhaps, agree.

Non-rechargeable primary battery such as button cells done of anode paired zinc (Zn) with liquid mercury (Hg) cathode using as salt bridge bases such as sodium hydrogen carbonate ($NaHCO_3$) sodium hydroxide (NaOH) or potassium hydroxide (KOH), magnesium dioxide (MnO_2) cathode-based that may be made using the ion sealed into the tetra-porphyritic ring of the chlorophyl extracted from the plant leaves, zinc-carbon dry cells done of zinc chloride ($ZnCl_2$) or starchy flour electrolyte between a zinc anode and a manganese (IV) oxide (MnO_2)-coated graphite porous rod cathode, lithium (Li) anode (T ≤ 45 °C, relatively higher voltage $\Delta V_{nominal}$ ≈ 3 V : 450 Wh/kg : 750 Wh/L) coupled with different cathodes (e.g. Li-air) even with solid glass (mostly silicon dioxide - SiO_2) electrolyte. More natural battery is the one done of soil called Earth batteries in which the elements attached to the clay cation exchanger acts in the hydrated electrolyte solution in storing energy which value depending on the potential of the ionized element in it (e.g., the leaves breaking down in the humus horizon V_C = + 30 V, V_{Mg} = -1.75 V => ΔV_{soil} = - 1.45 V) would certainly be relatively imprecise when the exact composition of the matter has not been investigated.

Rechargeable secondary cells firstly invented by Gaston Planté (40 Wh/kg, 90 Wh/L, 180 W/kg, <

350 cycles, $\Delta V_{nominal}$ = 2 V, - 35 < T 45 °C). In your souls, subtle body Atman Sharir, Vishuddha Chakra there is a biological lead (V_{Pb} = - 5 V) acid battery sealed in your vocal strings condensing the radiation of the planet Saturn ruled by the Paukert's law:

$$8 < t = \frac{C}{I^{1.3}} < 20 \, h$$

Where:
C = capacity at I = 1 A [F]
I = electricity [A]
t = time [h]

As the primary also the secondary battery can be done of plenty of different combinations between electrodes and electrolytes such as Zn-air, Ni-H$_2$ (60 Wh/kg, 60 kJ/L, 220 Wh/kg, p = 1200 psi, > 20k cycles), Al-ion (aluminium foil anode V_{Al} = - 0.8 V), Li-polymer, mostly, charged via the AC 220 V main electricity. Furthermore, the salty solution can be jellified in a gel done of sulfuric acid (H_2SO_4) and agglomerates of pyrogenic fumed silica (SiO_2 + 4 HCl).

Redox flow batteries are characterized by electrodes dissolved or colloidal solutions separated by a solid porous ion exchange membrane in which the voltage determined by the Nernst equation that can be used as rechargeable batteries or fuel cells (ΔV = 0.7 V, e.g. molten alkaline carbonate P = 100 MW, T = 600 °C, aqueous alkaline 10 < P < 200 kW, T < 80 °C, ethanol ion polymer membrane P < 140 mW/cm^2, 25 < T < 120 °C), these last electrochemical cells transduce chemical energy into reducer fuel (e.g., hydrogen H$_2$ boiling temperature T_{H2} = 20k K) and oxidiser (e.g., oxygen O$_2$) as in the Wormhole that gave origin to the whole Bubble Universe and in your antipodal subtle bodies, Nirvana (-) and Sthula (+) Sharir, Shahashrara and Mulhadhara chakras, lymphatic and bone anatomic system, respectively, surrounded by the heart magnetic field and divided by the cation exchange membrane of the present into two Einstein-Dirac conical halves filled of protons (H$^+$), which pairs can be placed onto a minimal energy β-snake path for the diffusion of the electricity or connected to external devices to recharge like an artificial intelligence e-embryo in need to be energetically supplied.

Superconductive magnetic energy storage (E = ½ L/I^2 = 3.5 TJ, 10 Wh/kg, 40 kJ/L, 100 MW/kg).

UltraBatteries are combination of lead acid batteries and ultracapacitors done of 50 % lead - 50% carbon anode, lead dioxide (PbO$_2$) cathode, sulfuric acid electrolyte (H$_2$SO$_4$).

https://www.intelligentliving.co/orange-peel-lithium-ion-batteries-new/
https://www.theengineer.co.uk/orange-peel-recycle-lithium-ion-batteries-ntu-singapore/
https://www.anthropocenemagazine.org/2020/09/orange-peels-could-help-recycle-old-lithium-batteries/

https://www.google.com/search?q=orange+peel+battery&rlz=1C1AVFC_enIT901IT901&oq=orange+peel+battery&aqs=chrome..69i57j35i39i362l7j35i39l2...7.5764j0j15&sourceid=chrome&ie=UTF-8

'Wormhole *in vitro*' © Copyright Antonio Silvestro, 2020

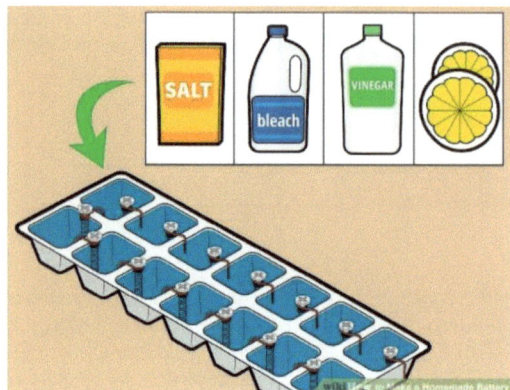

Figure 1 14 cells tray galvanized zinc nails enrolled by copper wires in water solution sodium dichloride (NaCl), sodium hydroxide (NaOH), acetic acid (CH3CO2H) or citric acid ($C_6H_8O_7$) generating a potential difference $\Delta V \approx 9$ V – Image source: https://www.wikihow.com/Make-a-Homemade-Battery

Biological cells have been used has model for glucose biobattery in which the cyclic monosaccharide hydrolysis is used for storing energy in its chemical covalent bonds.

Fruits like the *Rutaceae* lemons (*Citrus limoni*) and oranges (*Citrus sinensis*) containing citric acid, *Apiaceae* carrots (*Daucus carota*), and potatoes (*Solanum tuberosum*) tubers having phosphate (PO_4^{3-}) released by the Adenosine TriPhosphate (ATP) in which are placed an oxidizing cathode (+) and a reducing anode (-) like 1, 2, 5-euro cents coins [bronze = Cu (Cu^{2+}) 88 % + 12 % Sn] and a galvanized zinc (Zn^-) nail, connected via alligator clips or glued to black and red cables, respectively.

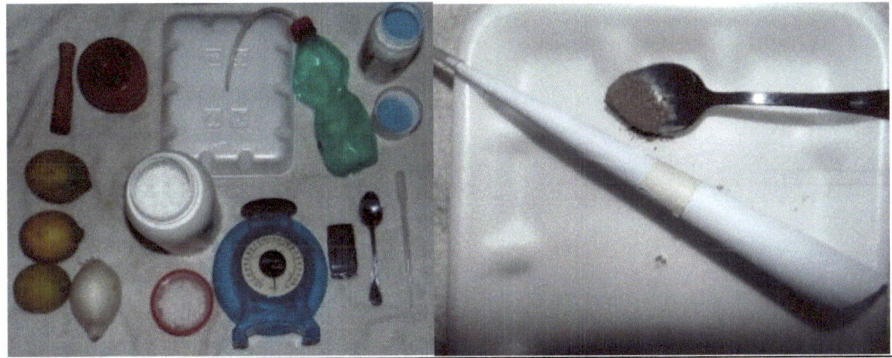

Figure 2 For obtaining a lemon battery with a volume of just V = 13 mL, five times smaller than the average lemon, containing $m_{citric\ acid}$ = 7 g suitable for triggering a potential difference ΔV = 5 V, a tea spoon of copper (fungicide 13 % Cu, m_{Cu} = 7.5 g) oxidised at 300 < T < 800 °C on an electric stove pre-heating it for t_0 = 5 min and later cooking in a metallic kettle for t_1 = 1 min changing colour from cyan to greyish. The copper dioxide (CuO_2) obtained would be inserted via a paper funnel in glass tube used for the hot steam distillation, dissolved in water (H_2O) and placed into ice box tray for being freezed (t = 4 h), double boiling for t = 5 min. Sulfuric acid (H_2SO_4), isolated from onion *Allium cepa* (m_{onion} = 750 g ≈ 15 onions, m_{H2SO4} = 0.5 g, 95 % 8 € + shipping/L https://www.letslab.it) evaporated from the smashed brattaeae half-onion ($m_{Allium\ cepa}$ = 25 g) with a wood mortar placed in a kettle Minerva oil bath (V_{oil} = 3 mL), condensed in the tube and collected into glass baker for t = 15 min, otherwise, made adoring Vulcan oxidising sulphur (S) powder heating it up and/or blowing air onto automatically via a compressor or manually via a syringe as Aeolus worshipper in a Athena double bowls oil bath, leaving the gaseous sulphur dioxide [$SO_{2(g)}$] reducing with liquid hydrogen peroxide

($H_2O_{2(l)}$), or generating it via Jupiter electrolysis supplying Cyclopic Direct Current (DC) lightenings (≈ 50 € https://www.aliexpress.com), or even got hydrolysing the sulphite (SO_2^{3-}) pellet of the Bacchus wine converted into vinegar (*'Dionysus/Bacchus – the spirit of enology and zymology'* KDP Amazon 1.92 € https://www.amazon.com/dp/B08F3V9SZD and Google Play Book Partner eBook 2.00 € https://play.google.com/store/books/details?id=26ElEAAAQBAJ), would be absorbed.

In parallel, squeeze 1 lemon (*Citrus limoni* m_{lemon} = 50 g, V_{lemon} = 50 mL; [$C_6H_8O_7$] = 1 g/mL) for isolating citric acid m_{C6H8O7} = 2.5 g (≈ 1 % of lemon), neutralizing it with the base sodium hydroxide (m_{NaOH} = 2.5 g) producing sodium citrate ($NaC_6H_8O_7$) that reacting with previously extracted sulfuric acid would make pure citric acid, that dissolved in concentration 10-times higher than the naturally present in the lemons, would permit obtaining affordable, reliable and long lasting 35 €/L ≈ 2000 batteries (ΔV = 5 V, 0.02 €/each), suitable for storing the energy of the planetary God Mercury contained in the Ceres vegetables – Image source: © Copyright Antonio Silvestro, 2020.

BATTERIES	
Pros	**Cons**
Power Density	Limited Cycle Life
Storage Capability	Voltage And Current Limitations
Better Leakage Current Than Capacitors	Long Charging Times
Constant Voltage That Can Be Turned Off And On	More Temperature Sensitive Than Capacitors
CAPACITORS	
Pros	**Cons**
Long Cycle Life	Low Specific Energy
High Load Currents	Linear Discharge Voltage
Short Charging Times	High Self-Discharge
Excellent Temperature Performance	High Cost Per Watt

Figure 3 Image source : https://www.machinedesign.com/automation-iiot/batteries-power-supplies/article/21831866/whats-the-difference-between-batteries-and-capacitors

The **soul** (Sanskrit: आत्मन् *Atman*) of the individuals, his the immortal expression that just *Chronus* can take away, firstly manifested in the groan of the baby, then in the voice of the adult emitted by the chanting vocal strings placed in the *Vishuddha Chakra* composed of fundamental pitch (50 < ν < 500 Hz) and harmonic frequency, the capacitor of the E-Homo that anatomically would be related to the endocrine thyroid gland secreting the thyroxin hormone (T_4) via the follicular cells increasing epinephrine and related sympathetic functions raising the heartbeat of the speaker.

Capacitors are passive electronic components ideated by Charles Pollak in the 1896, while supercapacitors by H. Becker in 1957 Metal Oxide Semiconductor Capacitors (MOSC) by M. M. Atalla and D. Kahng. The physical quantity that characterizes them is called capacitance which IUPAC system unit is the farads [F] and cam be calculate as follow:

$$C = \frac{Q}{\Delta V} = \frac{\varepsilon A}{d}$$

Where:

Q = Coulomb charge (+ or -) [C]

ΔV = voltage difference [V]

A = plates area [m²]
d = gap distance [m]
ε = permittivity [F/m]

Supercapacitors done of double chambers separated by a membrane (e.g., egg chorion) can store up to 100 times standard capacitors transferring and preserving charge for longer time compared to rechargeable batteries.

The Singularity certainly can to being in the present as it is going out under your eyes, in the now that is not anymore, a moment after, before where you were thinking what to do next or your past was following you, Hence, the sphere e with fuel that ignite the origin had to be placed in one or another half of the Wormhole sliding into the torus ring free in the widest space shielded by the electromagnetic field generated by itself. The conformational change between double Einstein-Dirac cone into torus electromagnet would have been manifesting the hypersurface of the present through which despite the size the dipolar Singularity could have been passed through when the permeable barrier would have been losing itself under its charming attraction. So, the nucleus would have been passed from cathode (+) to anode (-) when the surface would have been vanishing under the action of ribozymes breaking down the ribosomes catalytically, consequently, silencing the protein synthesis leaving metagenome free to diffuse bringing life to the other side of a polymeric transient insulating enclosure (ø = 50 Å) which aperture would have been triggered by the potential. In other words, inorganic composition of the electrolyte, electrodes, membrane between the empty space, but in a physical object would have been vary in the curving spacetime, the cylinder becoming bi-cone under the expansion of the Universe inflation foaming, itself changing into membrane [e.g., Nafion Proton Exchange membrane (PEM)] dividing the whole past and the whole future in two opposite volumes. For which you may understand that the cone itself has to be elastic, stretching when a relatively more rigid Singularity would have been aspired by the Black Hole (-) side for moving into the White Hole (+).

Dr. Masaro Emoto (Japanese: 江本勝) who claimed that human consciousness has an effect on the molecular structure of polluted water that could be cleaned through prayer and positive visualization on rice seeds: insulated (control) vs lauded. Actually, the H_2O crystals can change configuration according to the magnetic field surrounding them like the H-field of a human, reverse phenomena that could happen in the cytosols of the living cells when an external B-field is applied.

Helge von Koch snowflakes corpuscles developing from a triangle to a Judaic esa-vertex star can be seen according the A. Einstein Theory of General Relativity as AC continuous waves on which every Zeus electric arc has infinite length, because foldable into symmetric fractals infinite times, for which they can never be rectified by diodes or transistors into DC. This perfect geometry reveals how bending on its own is a natural awesome way for storing energy, peculiarly, electrons in the ground state from the light excitation that consequently would, after being reflected and diffracted, slowly melt the solid ice into liquid water freely the electrons in lenses reflecting and diffusing back light in the medium surrounding them. Infinite quantum energy is spontaneously stored in the perimeter of the irradiating Koch snowflakes along the circle of holiness that every meditative human has, but just a few know and have been institutionally declared able to manifest it. Without doubt, not a coincidence of the nature the correspondence of the planet Saturn with the cyan soul Vishuddha chakra, to remember the common mortals that their own halo pragmatically exists as the orbit of the planet ruled by *Chronus*, but just a bit farer looking in, up and around themselves with bright irradiating open eyes. In other words, the perimeter of the ice crystal is infinite, for this along it, the electromagnetic wave flow continuously in the folding Minkowski spacetime, manifesting its integer period just with the sublimation state of matter. Precisely, the perimeter is equal to:

$$\lim_{n \to \infty} P_n = N_n \cdot S_n = 3s \left(\frac{4}{3}\right)^n = \infty \text{ Triangle}$$

$$\lim_{n \to \infty} P_n = N_n \cdot S_n = 4a_s \left(\frac{5}{3}\right)^n = \infty \text{ Square}$$

Where:
N = side number
n = interaction value
S = original triangle side length
a_s = area of the original square

Despite the perimeter is a positive lemniscate, the area of the Koch snowflake accounts a finite value, infinite energy in a finite area equal to:

$$\lim_{n \to \infty} A_n = \lim_{n \to \infty} \frac{a_t}{5}\left(8 - \left(\frac{4}{9}\right)^n\right) = \frac{2s^2\sqrt{3}}{5} \text{ Triangle}$$

$$\lim_{n \to \infty} A_n = \lim_{n \to \infty} \frac{1}{5} + \frac{4}{5}\sum_{k=0}^{n}\left(\frac{5}{9}\right)^k = 2 \text{ Square}$$

Where:
a_t = area of the original triangle

A Chronus Hyper-Capacitor (HC) designed according the Koch snowflake geometry would slowly release energy to the supplied circuit into which has been plugged transmuting into less fragmented perimeter, on contrary, recharging it would subdivide its border more and more, in a theoretical never-ending capability to accumulate energy as continuous electromagnetic wave not differentiable.

Figure 4 Dr. Masaro Emoto experimental founding in which you may see the determination in choosing a way in the compassion, the isolation in the wisdom, the plurality in the sensing soul and beauty in the other crystals

'Wormhole *in vitro*' © Copyright Antonio Silvestro, 2020

(left), hexagonal ice crystals differentially gemmate on the six vertexes along a *'hydrotaxis'* pattern Koch snowflake (right) - Image source: https://www.tumblr.com/, https://www.naturesenergieshealth.com/health/water/dr-emoto-water/

A lightening rod, a relatively longer anode (+), placed on the top centre of the Chronus Super-Capacitor would catch the Zeus lightening, both natural and synthetic, and canalize their electricity inside. This irradiates in the direction of the peripheral hydrophobic walls of the cylinder containing alternate layers of insulating dielectric and conductive metal in a way that the electrons would continuously flow in an Alternate Current (AC). Crystal oscillator probes placed into Super-Capacitor can attract lightening far d = 50 km (http://lightningelectricity.com/).

Localizing elementary particles and their shah in the Big Bang wormhole

Big Bang nucleosynthesis (t_0 = 0 s and t_1 = 3 min) temperatures cooled from 10^{34} K to 10^9 K, and protons (H^+) and neutrons (n_0) collided to make deuterium (^2H).

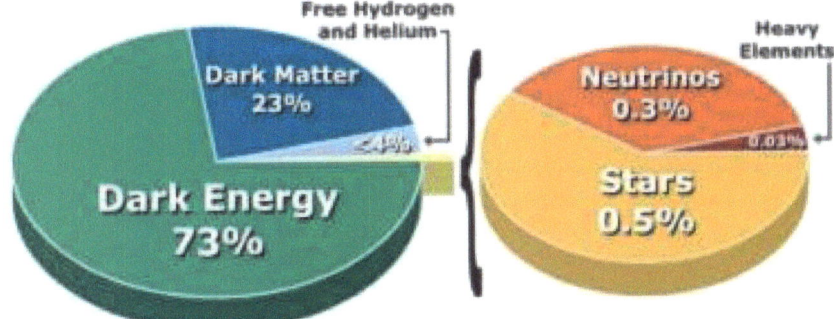

Figure 5 Universe compositive percentage cake charts.

Persistent, expandable and inter-active **matter** is any mathematic-physic-chemo-biological substance involved in the creation-destruction cycle, characterized by quantitative properties like mass and volume with five intra-changeable diverse main **states** of variable integrity and their corresponding physical objects such as plasma sparks, gas bubbles, condensate lenses, liquid droplets, solid particles, exhibiting wave-particle energy radio-transmissivity duality, field in which a mode host quantum excitations, relativistically anything that contributes to the stress-energy–momentum tensor of a system in the spacetime (about 70 % dark energy, 25 % dark matter, 5 % ordinary matter). Furthermore, matter from solid to liquid phase change in a nuance of *infra-states* before melting, passing through a glassy and a rubbery texture before flowing with a diverse viscosity phase. Glass-liquid transition and vitrification temperature are achieved super-heating or super-cooling a substrate, being at the antipode across the border between the amorphous and crystalline state.

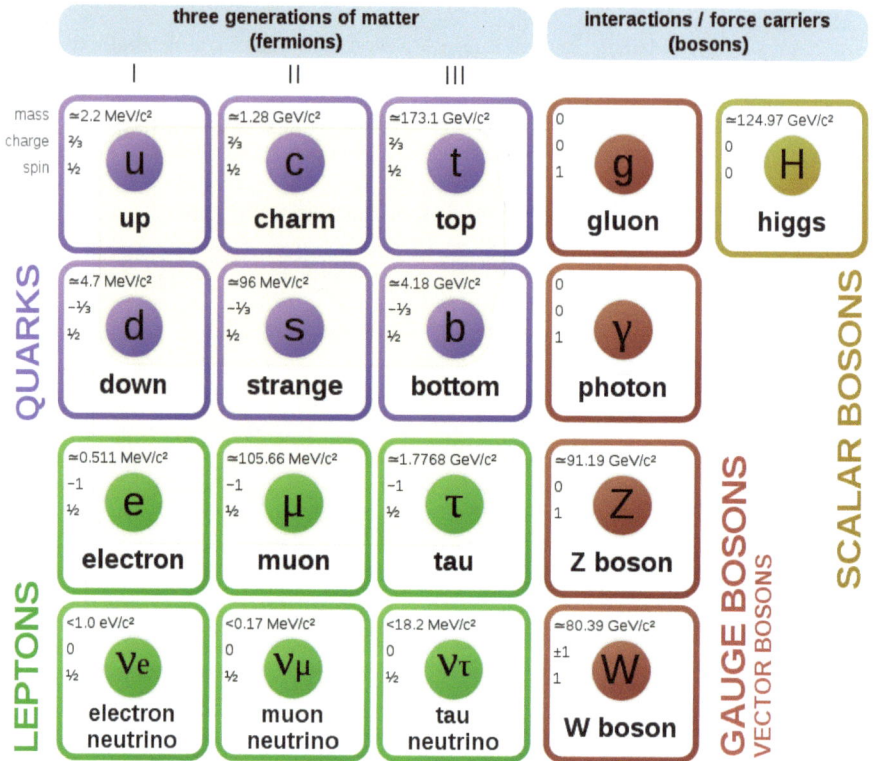

Figure 6 Standard Model Of Elementary Particles N.B. Sylvie Braibant, Giorgio Giacomelli and Maurizio Spurio in 2012 guessed the existence of bosons mediating gravity coined as graviton (G) (ISBN: 978-94-007-2463-1) - Image source: http://www.fnal.gov/.

The Sun continuously emits particles, mainly free **protons** (p ≈ H$^+$) through solar wind, and occasionally considerably augments its flow hugely with **Coronal Mass Ejections** (CME), colliding with the lightest **hadrons** and **muons** (µ$^+$ and µ$^-$). Hadrons are composite particles done of three strongly interacting fermions, one quark (baryons) or two mesons like *pion* (π^0, π^+ and π^-), *kaon* (K) or J/ψ, done of matter and antimatter, 12 fundamental fermionic opposite charged counterparts of the ordinary matter with which they mutual annihilate when colliding mainly releasing highly energetic ionizing radiation like *quarks* (q) and *anti-quarks* (-q). Within an atomic nucleus protons and neutrons are called **baryons** (from Greek: βαρύς, barys 'heavy'), triquarks particles [protons 2 up and 1 down (p = 2u + d), neutrons 1 up 2 down (n = 2d + u)] joined by *gluons*, while within a molecule the 3 freedom's degree (charge, spin and orbital) are associated with 3 *quasi-particles*, travelling unperturbed through the free-space: *holon*, *spinon* and *orbiton*. Furthermore, have been found temporary **hyperons** diffused by the Greek God Hyperion, such as the Λ, Σ, Ξ, and Ω particles, done of a variable number of strange quarks ($s_1 + s_2 +...+ s_x$, where x = numeral number N). Now would you believe your beautiful girl wondering on the Earth bring the Sun within you? Right there, in your empty space, super hole. But that's' not whole, hypotheticals **gravitons** and accelerant also could exist, they are related to gravity and Universe expansion close the its origin, your *Chitta*. Perceivable visible pinpoint, rapid and linear corpuscles flowing collide among themselves, and the internal 'complex walls' (e.g., metal$_{(s)}$, ocean$_{(l)}$, atmosphere$_{(g)}$, star$_{(p)}$, cosmic void) of the cavity of your seven subtle bodies *Sharir* in which they are content, marking their way via sinusoidal shahs of diverse colours according the PAR (e.g., blue: bad, induced by the other will; yellow-gold: very good, absolute, empty; white: neutral, objective, scientific).

Nuclear energy is proportional with volume ($E_n \propto V$), surface energy exchange is lesser than the one in the core of the cavity ($E_s < E_c$), the repulsion decrease the binding energy, balance in the amount of neutron and proton inside the nucleus determine higher isotopic energy, furthermore, an even number of particles is steadier, but constrict in a vinculum than the odd. **Atomic nucleus** (Latin: nucleus *'kernel'*) discovered by Ernest Rutherford in 1911 thanks to a Hans Geiger-Ernest Marsden experiment in which a spinning microscope measured the Anton Becquerel **decay** of the alpha α-particle (He-nuclei) scattering through a gold foil (Au) after which was noted that positively-charged protons (p^+) and chargeless **neutrons** (n_0) share a common envelope. The decay phenomena discovered by Enrico Fermi in 1934 conceal the transformation of the neutrons (q_n= 0 e⁻, spin$_n$ = 1/2, E_{n0} = 0.78 MeV) to protons (H^+) via W bosons (q_b= ± 1 e⁻, spin$_b$ = 1), releasing electrons (e⁻) and leptons antineutrinos (q_{ve} = 0 e⁻, spin$_{ve}$ = 1/2) in 15' [β-decay], while, inverse β-decay shows positrons, furthermore, 1 n^0/1000 can also emits γ-rays arising from the EM interaction with the proton β-decayed [γ–decay], moreover, $4/10^6$ protons can escape decay remaining attached among themselves 'two bodies' $H^+ + H^+ = H_2$, electrons can be caught by protons holes. Neutron background in the Earth fluid, atmosphere$_{(g)}$ and oceans$_{(l)}$, is due to muons (q_μ= - 1 e⁻, spin$_\mu$ = 1/2) produced by the cosmic rays. The n_0 Bequerellian also β-decaying releasing W boson, electron (e⁻) and antineutrino (ν) in two differently charged flavour quarks (up, charm, and top => electric charge = − 1/3 and down, strange, bottom => electric charge = + 2/3) involved in strong interactions with gluons. The n_0 have an inner stabilizing equilibrium between a positively charged core of radius $r_+ \approx$ 0.3 fm and a negatively charged shell's radius 0.3 < r. < 2 fm, so that the amount in their core is more populated in corpuscles than the envelope at fair absolute charge value. Free neutrons (n^0), influenced by the magnetic field (B), but not by the electric field (E), are hadrons made of 2 down - 1/3 and 1 up +2/3 quarks (baryons) held together by strong gluons, characterized by the following traits: mass $m_{n0} = 10^9$ eV/c^2 = 1.67· 10^{-27} kg, thermal energy $E_t = 4 \cdot 10^{-21}$ J can become colder in an equilibrating deuterium or helium solution in an enveloping cavity, fast energy $E_f = 1.6 \cdot 10^{-13}$ J = 1 meV, r^2 = 0.8 fm, spin -1/2, ℏ = h/2π = 6.6 · 10^{-34} J s/6.28 ~ 10^{-34} J s = 10^{-16} eV · s, E = 4.5 kcal = 1 J ~ 200 eV). The transition metal (d) network deflected the shell with two protons (2p^+) and two neutrons (2n_0), as the electron (e⁻) clouds were attracted by the electro-catalyst sheet while the protons reflected of 90 ° falling on the ground or flying in the sky, confirming that the human halo would protect protonation being itself already positively charged being the anthropomorphic Black Hole (+). The *Nirvana Sharir*, the empty subtle body is the gold bright that host electrons and perpendicularly repel protons giving them a direction according its position on the apical or occipital fontanella, closing or opening the static medium shipping into the healer trajectory the electricity to the father *Brahma* or rectified via the same pineal gland to the free *Brahman* to the all-Bubble Universe. The decay of the human during life is accomplished by the first path, the second would represent what should happen in the afterlife for which the isolation of an ephyphisis from a death body should clarify the issue.

Non-power reactors neutron-based that mimic the Sun are suitable for electricity, heat and propulsion production systems. Benchtop reactors can be done for reproducing α-decay, phenomena in which an atom releasing 2 H^+ becoming another one and β-decay using nuclides like beryllium (^{4}Be) and deuterium [Bq = 1 decay/s].

Theoretical algorithm for the conversion of high-frequency (ν) into current (I):

$$E_{neutron} = mc^2 \text{ [kg km}^2\text{/s}^2\text{]} = h\nu = 6.626 \times 10^{-34} \text{ kg m}^2\text{/s} \cdot 20 \cdot 10^7 \text{ s}^{-1} = 133 \cdot 10^{-27} \text{ kg m}^2\text{/s}^2$$

$$P_{neutron} = E_{neutron}/t = 133 \cdot 10^{-27} \text{ kg} \cdot \text{m}^2\text{/s}^3 = 133 \cdot 10^{-27} \text{ W} = V_{proton-electron}$$

$$I_{nuclide} = R \cdot V_{proton-electron}$$

Where:

R = resistance of the medium in the orbitals (s, p, d, f), each characterized by a different permittivity ε_r for storing differentially electrical energy.

While the planets, shielding and deflecting the charged particles, atoms, molecules with their magnetic field and even high molecular weight chemical compounds and neurotransmitter pulsing between telepathic humans in love and astronomical object would let the intensity of the radiations emitted increase with the latitude on Earth between them or the distance from the Sun in the *Solar System* (SS). Sudden widening flames of light from the Sun atmosphere (photo-/chromo-sphere and corona), the **Solar flares** red visible H-α particles decay from the 3rd to the 2nd spherical s-orbital, accelerating ($\Delta V = 10^9$ GeV) by the magnetic recombination with a total of kinetic, thermal and magnetic energy $E_{tot} = 10^{20}$ J, irradiated from the outer circumference of dark intense circular SunSpots of internal T_{SSi} = 5480 °C and external temperature T_{SSe} = 3450 °C, with their own magnetic fields, falling conically into ionosphere plasma medium (T = 10 · 10^6 K) causing bright polar greenish auroras on Earth (Solar flares A 10^{-7}< SI <10^{-4} W/m^2 X at Λ = 450 pm). Phenomena that you certainly, would visualize in your halo with Yoga practice in unity with the violet emitting - red absorbing *Shahasrara* of rebirth filling the orange emitting - indigo absorbing *Svadhistana* of endless life energy reflected onto the crystalline *Anahata chakra* omnidirectionally spreading greenish light of the love. The understanding would be behind the folding, that close the green Boreal aurora at the North pole of the planet Earth, with its centre should have been enveloped as a wet anus, scattering light while widening, showing its emptiness inside, the Ni-Fe magnetic shielding-attracting nucleus of the heart.

According the **Yukawa potential**, the smallest nuclei are the stablest, as the residual strong nuclear force that hold together quarks forming nucleons, decrease quickly with the square distance, leaving the thorium ($^{232}_{90}$Th) decay series formed radioisotope lead ($^{208}_{82}$Pb), being the largest known stable nucleus unchanging when α-β-γ-decaying. The **human subtle bodies spiral decay series** would follow an elliptical procession around each chakra, displacing horizontally in space and vertically in time, from the *Sthula* to the *Nirvana Sharir*, conjectures arising from the adoration of the Nordic God Thor associated with heart *Manas* and the Greek/Latin God Chronus/Saturn with the vocal string *Atman*. General formula of the *atomic decay series* along the periodic table:

$$A \ Z \ X_0 \rightarrow Z+1 \ A \ X_1 + e^- \ \nu_e$$

Where:
A = atomic number
Z = mass number
X_0 = initial element
X_1 = final element

Fermions obey Fermi-Dirac, while, **bosons** Bose-Einstein statistic. The rotation along the polar z-axis of the bosons is less perceivable as it is of 360° = 2π, than the half-integer front-back spin of the fermions. Boson (H, W and Z) are subatomic particles manifested when neutrons β-decay, experience quantum state and related superposition and Compton scattering when diffusing through the atmosphere medium. Bosons, entertaining particles regulators, are subdivided into: gauge drive forces damper, and intrinsic mass scalar bosons regulator. Everything exists, but in interaction, happening in four fundamental ways that make the bosons different among them: γ ElectroMagnetic, W and Z-weak, g strong and G gravitational. Peter **Higgs** leaded the discovery of its homonymous bosons at CERN using the **Large Hadron Collider** (LHC) (for more info about Kindle eBook 7.76 € Paperback 9.78 *'Olympus - the divine quadcopter'* © Copyright Antonio Silvestro, 2020 https://www.amazon.com/dp/B08GZ8MDWZ).

Would also fermions accomplish time-travel? Well, perhaps as complex aggregate excitation quantum leaping, transitioning till the boson emission as complex wavy signals, output of the present with the memory of the past in the future, in resonating groups, but lonely. Precisely, the accelerated massive complex subatomic particles coupled are the weak force responsible bosons W⁺ and W⁻ (spin s = 1, mass mW = ½ v · g = 80 GeV/c², where: v = Higgs vacuum expectation and g = gauge coupling) along a neutral Kaon oscillation width (Γ) with a wave function of the resonating energy equal to the relativistic **Gregory Breit-Eugene Wigner distribution**:

$$\Psi(E) = (2\sqrt{2}\ M\Gamma)/(\pi\sqrt{(M^2+M^2\ \hbar/\Gamma^2)})\ 1/((E^2-M^2)^2 + M^2\ (\hbar/\Gamma^2) \approx 2M\delta(E^2-M^2)$$

Where:
M = resonating mass [kg]
Γ = resonating antiferromagnetic chromium (Ch) silt width [m]
ℏ = h/(2π) = 1.054 · 10⁻³⁴ J · s

The omnipresent **neutrinos** interact with lighter leptons emitting Cherenkov radiation exchanging Z-boson in neutral currents, on contrary, colliding with heavier leptons swaps W-boson in charged flows that let transmute the targeted particles too. They can be detected via scintillators, radio-chemical Bruno Pontecorvo reactions, abysm and radio ice Cherenkov detectors, tracking calorimeters and extremely small devices based on the coherent neutral current neutrino-nucleus elastic scattering. Neutrino detectors can be used for predicting the happening of a new *SuperNovas*.

J. J. Thomson suggested that the e⁻, four order bigger than the protons (H$_\varnothing$ = 156 < electron diameter < U$_\varnothing$ = 53 pm, H$_\varnothing$ = 1.8 < proton diameter < U$_\varnothing$ = 12 fm), hence, indirectly that being the neutron proportional to the protons, the nucleus is almost empty of elementary particles, are randomly scattered by the proton sphere. **Electrons** according their three DOF independent variables on which depend their probabilistic distribution (∝ Ψ) are distinguished in quasi-particles: holons (charge), spinous (spin), orbitons (orbital).

Cyclotrons, electron in orbital motion, mentioned by Sir. Albert Einstein, released from nuclides decaying in the horizon O, could be considered as 'electrons guiding evolution' when flowing in the seven human subtle bodies, seeing random movement governed by the Lenz's law, for example, canalized from the *Bhava Sharir* Earth ground into the *Sthula Sharir* solenoidal coccyx, insulated within the calcium bone networks, of a seated meditator. Nevertheless, the **cyclotron energy** in one ascending dimension congruent with the inner magnetic field (B):

$$E_c = T(R/N) = (8\pi/L^3)(R/N) = [8\pi/(x \cdot y \cdot z)](R/N) = 2/3\ E_k + 1/3\ E_p$$

Where :
T = absolue temperature [K]
R = Universal gas constant = 8.314 J / mol
N = reagent mass coefficient, coming from M. Planck and J. C. Maxwell through A. Einstein = $(\beta/\alpha)(8\pi R/L^3) = (\alpha/\beta) = 6.17 \cdot 10^{23}$ [gram equivalent = mass number (A)/valence, e.g. hydrogen H 1.008 u/1 = 1.008, Og 294 u/118 = 2.49] N.B. N ≈ NA Avogadro's number $6.022140857 \cdot 10^{23}$/mol.

8 π depends on the octet valence being the maximum translocation path accomplished by an electron in the outer s-orbital (just spherical orbitals can be managed as spherical N.H.D. Bohr orbits).

L = product of the three spatial dimensions of the radiation absorber (e.g., photosynthetic pigment, Black Hole, etc.)

Biot-Savart law for one solenoid loop:

$$B = \mu_0 I/4\pi \int (dl\ \hat{r})/r^2 = 1.26 \cdot 10^{-6} \cdot (R \cdot V_{proton-electron})/12.56 = \int (dl\ \hat{r})/6.4\ fm$$

Where:
μ_0 = vacuum permeability = inductance/ length = $1.26\ 10^{-6}$ H/m
I = electric current [A]
l = length solenoid = f_{xy}
r = loop radius [m]

Weakly interacting **quasi-particles** like phonons in crystal Bravais lattice, excitons bonding electrons and holes, plasmons of plasma, polaritons bridge between ElectroMagnetic radiation and dipoles carrying it, precisely, between photons and electrons, polarons involved in the adiabatic electron current typical of the phase transition (gaseous -> liquid -> solid) and magnons collectively electrons spinning in an excitation field coherent with its domain. Dark matter could see its light in the **Gravitationally** or **Weakly Interacting Massive Particles** (G-or W-IMP).

Wave functions vector for discrete particles and quasi-particles with any spin (s = - 1, - 1/2, 0, + 1/2 or + 1) and their continuous corresponding radiations captured by Black Holes, within in the Hilbert space can be described by the following formula:

$$|\psi(t)\rangle = \sum_{s_z} \int d^3\ r_s\ \psi(r_s, s_z, t)|r, s_z\rangle$$

Where:
r_s = Schwarzschild radius = $2\ G\ (M_\circ/c^2) = 2\ G\ (E) \approx 2.95\ M/M_{Sun}$ km
s_z = 'spin projection quantum number along the z-axis', a discrete variable equal to the spin and its opposite value for any massive particles.

Fluctuation's three-dimensionality was discovered by Leonhard Euler, deducted from the spherical waves:

$$\Psi(\vec{r},t) = \int_{-\infty}^{+\infty} \Psi(\vec{r},\omega)\, e^{-i\omega t} d\omega$$

The plane on which particles lay can vibrate with them at the unison when the frequency of its oscillation in enough small to be compensate by the mass of the corpuscles, otherwise, the bodies on it roll and jump. Condenser plates in an engine of vehicle riding on countryside road while an earthquake would be an example of Universe in which the common condition is the plane vibration, the perturbative waves resonates among them from the smallest to the biggest cosmos chaotically, making the core system with highest entropy values. Any of the **plane waves** can be obtained with the following formula, for which each of the sub-unitary packets travel coherently within the frequency domain (2D: circle, ellipsis, parabola or hyperbola, 3D: sphere, ellipsoid, paraboloid, and hyperboloid): $f(x) = A\, e^{\pm ikx}$, where: A = amplitude [m^2], i = imaginary unit, k = wavenumber [m^{-1}], x = distance [m].

Nuclei can be spherical (s), prolate (p), oblate ellipsoids (e) or toroid (p, d, f) or a combination of their shapes depending on the electron turning around on orbits as modelled by Niels Bohr in an indeterminate time gap unavoidably reaching the center, waves of probabilities surrounding the nucleus as defined by Max Born, "At quantum level all the basic laws of common sense are violated: electrons can disappear and reappear elsewhere, and electrons can be many places at the same time " (*'The physics of the impossible'* by Michio Kaku 加來道雄). According to the quantum theory, electrons vibrating at the unison z-eigenstate in a coherence domain even if they are separated by infinite distance, before the scientist make the measurement, their imagines spins neither up nor down, but exists spinning both up and down as the Schrodinger cat both alive and death, becoming detectable at the instant during which the complex wave function of the observer collides onto them assessing their localization. Finally, e$^-$ could be described as probabilistic roto-traslatory negatively charged sub-atomic particle in stationary 3D **orbitals**, theoretically determinable by the 'freezing' tensor formula dividing the rotation along its main vertical axis (spin) and the (counter-)clockwise (+ or -) translation along the perimeter of the 2D orbit, pragmatically, catching them with a network of positively charged holes like the Nafion PEM highlighting their position at time of their passage into it.

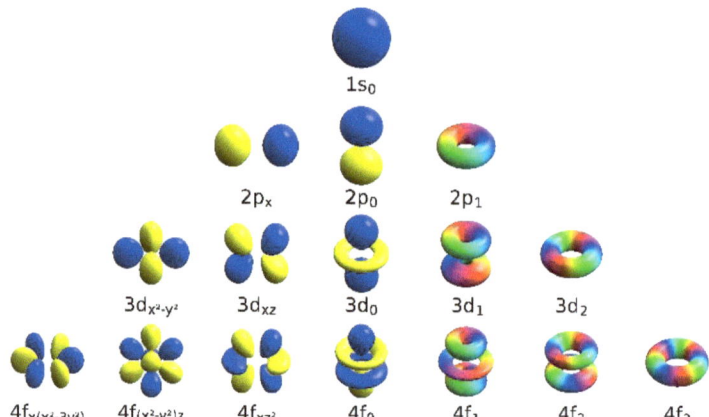

Figure 7 *Electron duality vs unity* for occupying the same transitory space in superposition's ± m (left) or symmetrically rotating along the vertical z-axis m_z-eigenstates (right). Atomic single electron n = l + 1, where: radial node = l = 0 (s), 1 (p), 2 (d), 3 (f). - Image source: https://commons.wikimedia.org/wiki/User:Geek3

The eigenstate would be manifested, just after the superposition in less probable events in which the induction of the magnetic field according the Faraday law is the fundamental onto which the resonating natural frequency of the electron carrying the wave of the *Anahata* chakra is. Fight or love, but alone divided, centered and balanced with no heart measuring the **nuclear radius** with the following formula:

$$r = A^{1/3} \, r_0$$

Where:
A = Z + N = proton + neutron = atomic mass number
r_0 = 1.25 fm

Well, certainly, that's who love for fight and flight too, but a part of it the mathematical entity that place the divergence among the motion nuclei describing is the at the *Del* or *nabla* operator (∇) elucidating the gradient of a function in Euclidean space $\nabla < 0$ = colliding, $\nabla > 0$ = escaping, $\nabla = 0$ divergent. Its square, is called Laplace (∇^2), symbolize the gradient of differential step unit's area, is present in the **Klein-Gordon-Fock relativistic wave equation** suitable for determine the destiny in spacetime of the electrons:

$$\frac{1}{c^2}\frac{\delta^2}{\delta t^2}\psi - \nabla^2\psi + \frac{m^2 c^2}{\hbar^2}\psi = 0$$

Where:
Planck constant ℏ (h-bar) = $1.05 \cdot 10^{-34}$ J · s

Probable distribution (P_d) of a corpuscle of otherwise uncertain localization can be represented by a wave function Ψ (t) of amplitude (A) defined by a complex number:

$$A = (a + bi)$$

Where:

a, b = real number
I = imaginary number ($i^2 = -1$)]

Complex conjugate of $A = B = (a - bi)$

Probability of distribution of a particle x in the range from a to b (e.g., e⁻ 1s σ-bond H_2, Black-White Hole) in the 3D space Hilbert or in the infinite normalized joint that localize the existence of the particles:

$$P_d\ a \leq x \leq b\ (t) = \int_a^b dx\ |\Psi(r, \theta, \Phi)|^2 = 1$$

With high probability, most of the dark matter is sealed in the past before the Big Bang, and for knowing its exact nature would be need to turn the head of the observer, in the White hole, more than in the Black, but certainly, that's dictate to the natural behaviour of the humans overcoming the selection for thriving going ahead without turning the face back into their powerful gonads.

Perhaps, the most famous example of quantum weirdness is the **'Schrödinger's cat'**, a thought experiment devised by the Austrian physicist Erwin Schrödinger in 1935 who imagined how a cat placed in a box together with a radioactive isotope and a poisoning ampoule releasing its content due to the sub-atomic particle decay. The *Felis catus* is said to exist in a superposition state, according the odd laws of quantum mechanics, both dead and alive at least until the box is opened and its content observed depending on the transition decay of the radioisotope constructively or destructively with the administered venom. We humans and all the other sentient being exists in the domain of the complex numbers, where interferences make us be. Schrödinger wave functions in the organic molecules are stronger than the ones involving electrons coming from inorganic atoms, as the electron-holes pair are better localized flowing in monomer relatively more stable.

Superimpose between two fringes, we bring in our silent Manas, *Gemini* the constellation that is both dead or alive in the space of our happy heart which funny current could be measured using **electrometers**, invented by Theodor Wulf in 1909, are devices suitable for measuring the electric charge (q) and/or tension (ΔV) deriving from their differential flow through a complex empty limit to overcome like the brane of the Bubble Universe, growth medium, cellular membrane. Is the void limitless? Certainly, not. Just full of elementary particles. Sources of **radioactivity** are ubiquitarious as the subatomic particles that generate them, this can be on ground as discovered by Anton Becquerel, underwater as concluded by Domenico Pacini and in the atmosphere as deduct by Victor Hesse, thrice researcher working in the beginning of the XX century, but with a different richness that increase exponentially with the latitude. It is guessable that within the humans, elementary irradiating corpuscles would be more abundant in the *Nervous Central System* (NCS), than the reproductive and digestive system as in the *Solar System* (SS) on the planet Uranus, followed by Mercury, finally, Earth. The amount of chaotic sub-atomic particles influences the **enthalpy rate** measured for the first time by Werner Heinrich Gustav Kolhörster, Noble Prize in Physics in 1936. Chlorine-34 (^{34}Cl) 32 min < *radioisotope half-lifetime* > Beryllium (^{10}Be) million year. Radiation come from our current planet, but also from other like Mars as evaluated in 2012 by the **Radiation Assessment Detectors** (RAD). At present, the most intense source of nuclei disintegration are the

'Cosmic Rays' defined by Robert Millikan in 1920, precisely, referring to the ionizing radiation primarily due to protons and secondarily to unstable decaying atomic nuclei containing leptons (electrons and muons like pions and kaons) and boson photons that would condense in the *Manas* crystals forming phonons and converting magnet in magnons, while, and in the *Bhava* water becoming *Dipole Wave Quanta* (DWQ), released by supernovas and *Active Galactic Nuclei* (AGN), that coming from the interplanetary space, outside the *Solar System* (SS) or even the *Milky Way galaxy* (MW) reaches the Earth's atmosphere colliding with its main molecules and atoms like oxygen (O_2) and nitrogen (N_2) ionizing them and inducing radioisotopes production like 14-carbon (^{14}C). Among this, the **Ultra-High-Energy Cosmic Rays** (UHECRs) have been observed to approach tensions of $\Delta V = 3 \times 10^{20}$ eV and energy value of E = 160 J (3-fold the OMG particle!). **Spallation X-Process** is the productive process that use the cosmic rays for bombarding nuclei and breaking-up them into several particles that remain hanged and vibrational, hence, suitable for following nucleosynthesis. A pragmatic application of this physio-chemical operation is the *plastic recycle* as experimented by R. Fleisher, P. Buford Prince and Robert M. Walker on *PolyCarbonate* (PC) balloons samples.

Magnification of distant astronomical object is accomplished by telescopes refracting lenses, reflecting mirrors and catadioptric combining mirrors and lenses. Galactic in South Hemisphere and even Extra-galactic in the North Hemisphere γ-rays' reflective detectors such as the aromatic grid parabolic ground-based Imaging Atmospheric Cherenkov' s **Telescopes** (IACT) are aimed to generate tensions till $\Delta V = 300$ TeV thanks to reception of the highest ionizing radiation existing (https://www.cta-observatory.org/) measurable with a *Geiger counter*. Edwin Hubble **Space Telescope** Cassegrain Ritchey-Chrétien edited design reflective bi-hyperbolic diffraction limited mirror (A = 4.5 m^2) polished to an accuracy of λ = 10 nm, Al and MgF coated, suitable for UV, PAR, near IF, propelled along a *Low Earth Orbit* (LEO) at altitude h = 540 km by a expelling exhaust fuel rocket as classically described by Hermann Oberth, together with the others telescopes have the angular resolution and the wide-spectra, due to the atmosphere scattering avoidance, advantages compared to the terrestrial astronomical observatories. The successor James Webb **Space** (https://www.jwst.nasa.gov/) (A = 25 m^2) that would make human understand the conditions of the Universe $200 \cdot 10^9$ years ago ($\lambda_{human\ eye}$ = 0.6 μm and aperture human eye = 1 cm, while, λ_{JWST} = 28.5 μm and aperture human eye = 6.5 m) is done of Red ^{49}Be frame with 79196Au anodized mirrors within (18 hexagonal aromatic rings: O, 6 along first inner-ring more 12 along the second outer-ring), five-layer polyimide sunshield coated by Si and Al, vacuum cooled with cryocoolers better than nitrogen (N_2) and helium (He) till T = - 220 °C permitting observations of a low energy spectra till mid-*InfraRed* (IR), hence, covering a widest range of planetary emitted radiations. Radiometric sensitivity for space telescope would reach in the next feature the high values of λ = 1 mm suitable for investigating the genesis of the Universe across the Big Bang. In 1948, observations with nuclear emulsions carried by balloons to near the top of the atmosphere showed that approximately 10 % of the primaries are helium (He). According the **Hubble – Lemaître law** astronomical objects ≈ 10 Mpc are subjected to spectrum's frequency Doppler red shift (z = 0.2) which remote control can be done via transmitters Earth-based. Influencing their motion as for the planets the roto-traslating around the core planetary star. ElectroMagnetic radiation stellar spectrum emitted from celestial objects characterized by Doppler shift due to the relative motion, away (red) or close (blue), of the wave, permits determine physio-chemical properties such as composition, temperature, density, distance, luminosity. Orbital telescope is generally used for detecting UV, while, antennas for noise detection of radio frequencies. The first that splitted light with a prism was Isaac Newton followed by Joseph von Fraunhofer that commercialised high-quality refracting telescope based on the amphitryon refined glass system capable to detect 574 dark lines/mm in the continuous light spectrum. J. S. Plaskett invented grating parallel mirrors that incremented the light focalization in a determined wavelenght range and the yield by tilting the assembled reflectors until 1k lines/mm, limit overcame by volume phase dichromatic gel holographic Bragg diffractive interferometers reaching 6 k lines/mm.

Stars Elements	Wavelenght [nm]
O_2	899-628
H	656-410
Na	589
He	588
Hg	546
Fe	527-302
Mg	517
Ca	430-393
Ti	336
Ni	299

Interstellar medium composition: $H_{(g)}$, $He_{(g)}$, $O_{2(p)}$, $C_{(s)}$, $SiO_{4(s)}$, $H_2O_{(s)}$.

Cloud created by Wilson and **bubble chambers** invented by Donald A. Glaser awarded of Nobel Prize of Physics in 1960 were used for particle investigation. In the second type $H_{2(l)}$ medium were boiled by surrounding heated solenoids increase its volume pushing down the piston within the cylinder, but without changing state of matter to gaseous (super-heated). Particles in them that succeed in overcoming the shielding magnetic field via differential helical path due to diverse charge/mass ratio (q/m) and velocity (v), maintain hydrogen in a meta-stable liquid phase (e.g. W and Z bosons inducer of weak neutral currents), and are photographed. Currently, the aforementioned physio-chemical medium-based chamber for sub-atomic particles identifications are supplanted by electronic devices like solid-state transparent optical semi-conductors (e.g., ^{14}Si, ^{32}Ge, $^{48}Cd^{52}Te$, $^{14}Si^6C$, $^{31}Ga^{33}As$) **Multi-Wires Proportional Chambers** (MWPC), invented by Georges Charpak in 1968 (originally 10^3 particles/s), Noble Prize for Physics in 1992. In this resistive electro-plates [reference anode (-) and counter cathode (+)] condense the filled gaseous solution done of an inert noble proportionally sensitive scintillating element along the wire like Argon (^{18}Ar) and a nuclide source like the alkane Methane (CH_4) pulsing plasma currents formed from the ionization of this last, due to the atomic carbon and hydrogen collisions that strip-out their valence and freely their co-valence electrons in a dense neutral tracked current amplifiable for its digital charting. Electron-dense orbitals would permit the determination of the most probable e^- localization (electron-density), point per point, on electricity map injecting photons of determined wavelenght in the detector for labelling with fluorescence (emission after absorption) the e^- pathways. The less energetic visible radiation (λ = 380 nm) would knock-out the covalent e^- (CH = 1.09 Å and ECH = 413 kJ/mol), while, the higher P.A.R. (λ = 700) releasing the valence negative charged leptons. Actually, effective accelerating drift and emission (transmission) of the lepton's electrons (e^-) in liquid phases via *ElectroLuminescence* (EL) in thin gap of the **Thick Gas Electron Multipliers** (THGEMs) detectors have been developed in laboratory for dark matter research (80 % of total matter) (A. Bondar et. al, 2019), this suitable also for Townsend avalanche signal amplification in gaseous medium.

Cylindrical **Time Projection Chambers** (TPC) invented by David R. Nygren in 1970, use ionized fluids media in ElectroMagnetic Fields for determining the corpuscles trajectory and interaction (*'European Council for Nuclear Research'* https://home.cern/). Collection and magnetic induction perpendicular wires make an anodic (+) network hosting the drifted electrons (e⁻) stripped out from the charged particle injected, polarized from the random Fermi thermal velocity due to the collisions, exactly, difference between highest and lowest quantum state at temperature T = 0 K, hence, determinable *velocity* (v) under electric field (E) [V/m] proportional to the current (I). Precisely, the **drift velocity** is equal to:

$$v = \mu E = \frac{j}{nq} = \frac{m\,\sigma\Delta V}{\rho_e f l} = \frac{I}{n A q} [m/s]$$

Where:
μ = variable electron mobility through a conductive material [m²/ (V · s)]
$j = I \cdot (x / t)$ = current density [A · m/s]
$n = M/ V^2$ = charge-carrier number density [mol/L²]
q = electric charge [C]
$q_e = e$ = unitary charge positive (p or H⁺) or negative (e⁻) [C]
m = mass [kg]
ρ = conductor density [kg · m³]
σ = electric conductivity [S/m]
f = free electrons per atom [e⁻/a]
l = conductor length [m]
I = electric current [A]
A = area [m²]

Nuclear destruction vs creation : H-Bomb vs Tokomak

And God said, "Let there be light," and there was light. God saw that the light was good, and he separated the light from the darkness. God called the light 'day', and the darkness 'night'. And there was evening, and there was morning — the first day (Bible, Genesis 3-5).

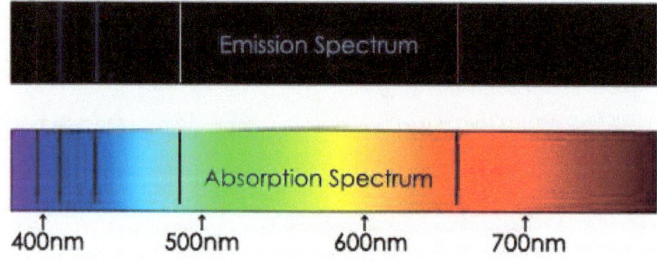

Figure 8 Hydrogen spectrum

Hydrogen (₁H, energy electron at vacuum ground state E_{e^-} = -14 eV) firstly generated during the Big Bang recombination epoch, is colourless, odourless, tasteless fuel that account the 75 % baryonic mass of the Bubble Universe auto-ignite, usually inert being characterize by a (H₂ : H-H bond dissociation energy into E = 436 kJ/mol) at higher temperature of T = 500 °C, while, when filling your solar plexus *Manipura* chakra and *Shukshma* sharir spontaneously exploding in the chloridic acid (HCl) stomach of the Greek/Latin God Hephaestus/Volcano at temperature T = 37.5 °C when the sparks of the father Zeus/Jupiter ionize the gases in the circulatory vessels carrying O_2 or CO_2 into hot discharging plasma filaments igniting the medium with flames (Latin: *flamma*), visible part of a fire due to exothermic chemical reaction. The core of the Solar System is characterized by nuclear reactions that involve the conversion of hydrogen in helium [$_1^1H_2$ = 2 (1 proton + 1 electron) into $_2^4He$ = 2 proton + 2 electrons + 2 neutrons https://www.ptable.com/]. The invisible and dark zones of flame's emitter such as a candle, resembling coalescent black holes confined by blue light (T_b ≈ 800 °C), are characterized by the average temperature T_i ≈ 600 °C and T_d ≈ 1000 °C, respectively, beigns the first volume not directly involved in the phenomena, but supporting it, as the past Einstein light cone, and the second a contrasting zone that stands out the stellar brightness of the reddish-orangish-yellowish incomplete combusted hydrocarbon soot of the luminous zone (T_l ≈ 1200 °C), itself enveloped by a water vapour of the non-luminous veil (T ≈ 1400 °C). In a noble inert atmosphere, with no oxidizers (e.g., oxygen O_2, hydrogen H_2), occur the pyrolysis of the fuelling wax releasing heat as soon it become vaporized. In excess of oxygen, the complete combustion on the fuel, without waste and related black body-radiating soot, produce the cooler blue (λ ≈ 565 nm) flames than the yellow, in which molecular radicals, especially methylidyne (CH) and diatomic carbon (C_2) band emission. Pure hydrogen- oxygen (H_2-O_2) flames emit ultraviolet light (photon γ-ray $λ_{UV}$ = 90 nm). The first element of the periodic table can be produced by water electrolysis, methane steam reforming and pyrolysis, alkali and alkaline earths metal-acid reaction, thermochemical cycles. It can be used for Gas Chromatography (GC), to synthetize methanol, ammonia and cyclohexane or for hydrogenate edible lipids.

Figure 9 The simulation of Jupiter lightening on the Poseidon surfaces with electrolysis of a potassium hydroxide in glass bottle (m_{KOH} = 7 g, V_{H2O} = 700 mL => [KOH] = 0.1 g/L) with photoautotroph bacterium-plankton *Cyanobacteria Arthrospira spp*. fed once with $V_{NPK\ 505}$ = 1 mL and placed at direct Helios sunlight

'Wormhole *in vitro*' © Copyright Antonio Silvestro, 2020

('*Fluidoponics*' eBook 3.57 € https://www.amazon.com/dp/B08DL9B488), suppling via graphite (C sp$_2$) electrodes, reducing acid cathode (+) oxygen (O_2) and oxidizing alkaline anode (-) hydrogen (H_2), the reactive chamber with an AA battery (ΔV = + 1.5 V), storing the bubbling hydrogen into a smaller tank (V = 100 mL) from which it would be released via a syringe needle - Image source: © Copyright Antonio Silvestro, 2021.

Thermonuclear weapon based on fission release X-ray that heats up the cool chamber increasing its internal temperature, according to the universal gas equation, compressing the fuel inside (p ≤ 1.4 x 10^9 atm), leaving it fuses yielding energy E = 210 TJ sealed into a tamper radioactive Uranium barrier controlling the imploded ionizing radiation. In the Teller–Ulam chain configuration, the fissing implosion would boost the fusion of the fuel (50 % deuterium / 50 % tritium) releasing the neutrons (n^0) is excess when heated and compress. While the Tsar Russian Bomb is characterized by three fusion stages in which the energy is transferred as compressed heated radiation.

Spheric I fission chamber: Tritium core, Uranium-235 or Plutonium, surrounded by vacuum, absorbing Uranium-238, high explosive prism lenses canalizing the radiation onto the walls leaving it partially absorbed spreading in all directions inside the cell and over a thresh-hold emitted out into the chamber II were fusion occur as electromagnetic radiation and neutron flux superheating and expanding gases ionizing into plasma. Two chambers enclosed in a reflective hallow space *holraum*, which X- and γ-rays' photons opaque walls are in a radioactive equilibrium with the radiant energy within the I column. The radiation equiparable to the black-body one, escape through the ablating I walls and is canalized via the Supercritical Head Space where neutrons are polarized to light, low density Expanded PolyStyrene (EPS) foam, aerogel, or fogbank, but almost uncompressible ablation delayer, becoming hot plasma onto the II walls where pushed inside implode. The fission chamber II is done of golden coated Uranium-238 tamper layer containing lithium 6-deuteride dry fuel compressed, bombarded with neutrons splits into tritium and bosonic alpha particles (2 protons and 2 neutrons identical to He nucleus, He missing its two electrons), undergoes fusion releasing energy, reaching the plutonium ($_{94}$Pu) spark-plug hollow core and fissing itself.

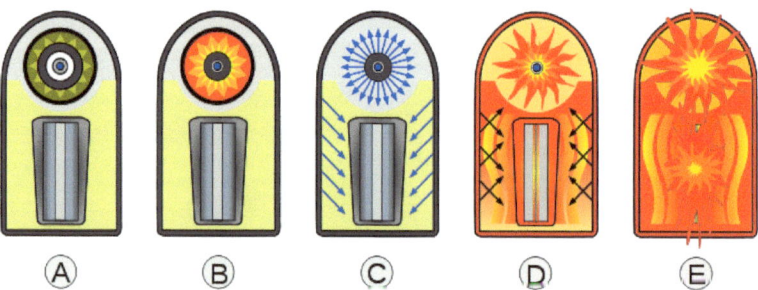

Figure 10 Foam plasma firing sequence: A. spherical fission bomb in *sensu stricto* (I), parallelepipedal fusion fuel Pu spark plug (II) enveloped by EPS suspension, B. Compression, C. X-rays scattering, D. EPS from gaseous to plasma under compressing vacuum (II) fissing the Pu, E. Li-6d compressed and heats the Pu fuel, that release 3H$^+$ and α-particles, fusing culminating in a fissing U-235 tamper neutron flux

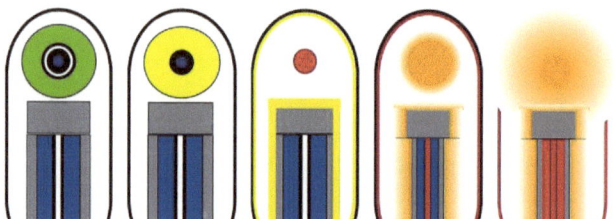

Figure 11 Ablation: 1. I nested spheres fission II cylinder fusion 2. Pu fissed 3. II over heated, X and γ-rays!0 irradiated against the internal holraum and contiguous external II walls 4. II walls removed, pushing U-235, Li-6d and Pu inwards, fissing the Pu spark 5. Li-6d fusion.

Omitting the last fission stage replacing Uranium tamper with dense Cronus lead (plumbum $_{82}Pb$), the explosion in the ion half reduce in intensity and the radioactive propelled into the atmosphere (nuclear fallout) is included lower than half. Perhaps, using dissolution of lead into Zeus tin ($_{50}Sn$) would reduce the environmental toxicity for a H-generator design.

The energy is deposited within about one X-ray optical thickness, the Napierian logarithm of the ratio of incident to transmitted radiant flux (Φ) [W = J/s], the partial derivate of the radiant energy emitted, reflected, transmitter or received on time, and its spectra ($\Phi_{x,v}$ or $\Phi_{x,\lambda}$) [W/m or W/Hz]. The physic optical depth ($\ln\Phi_i/\Phi_t$) is the analogue to the chemistry decadic absorbance, otherwise, $\log_{10}$ based. Vaporized pusher gas expansion and ionization into plasma velocity up to v = 570 km/s, implosion fission pressure (p), radiation p = 1.4 Gatm, plasma p = 7.5 Gatm, ablation p = 64 Gatm.

The elastic H-bomb would release lethal energy in one only pulse, but if the compression due to the heat would be subdivided into n-steps it could eject differential energy in storing batteries to it connected for being later utilized for charging compatible electronic devises. Perhaps, channeling the energy outgoing from a Teller–Ulam H-bomb, derived from fission and following fusion of the nuclei, shaping the wave through optical lenses transmission system packaging with tiny steps to reach a high-voltage, them in coherent volumes – in Chronus "differential hyper capacitors", would healthily regenerate the spirit of life in a sustainable chain reaction for a smart-grid distribution. Cooling hydrogen under vacuum should let its nucleus to merge, perhaps, forming intermediary isotopes, deuterium and tritium, later helium canalized into a storage fission chamber where the X-rays emitted are stored in a radiation reflective case. Assuming the frequency as sinusoidal turning around the radius axis, could be calculated this last based on the X-rays (1 nm < wavelength λ < 1 pm, 100 eV < photon energy Eγ <100 keV) emitted from I column, designing a holraum with a phallic shape that would ablate at I column side containing the high-energy radiation for regulating the ejection on demand.

Hydrogen Internal Combustion Engine Vehicle (HICEV), battery or (super)capacitors, modified gasoline-powered with gas instead than liquid fuel, harden valves, connection rods, non-platinum tipped spark plugs, higher voltage ignition coil, fuel injectors, larger permit decrease an average of 1.5-fold the current costs and an energy output major of approximatively 15 % BMW Hydrogen 7 v = 301 km/h. Air: H_2-fuel 34 : 1. Low emission combustion: $2\ H_2 + O \rightarrow 2\ H_2O$. Hydrogen Fuel Cell Vehicles (HFCV), electrochemical conversion, rather than combustion, to generate electricity for power engine, using O_2 and H_2 compressed, emitting water and heat.

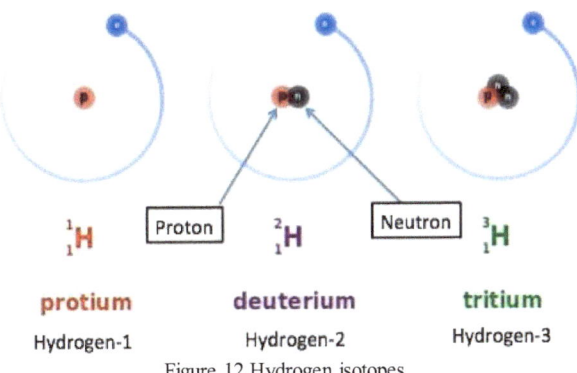

Figure 12 Hydrogen isotopes.

The values to asses and improve to the utmost in the 2^n chambers domestic bench-top in a series H_2-

Bioreactor are: Hydrogen Recovery (optimum ≥ 90%), Hydrogen Production Rate (HPR), Coulombic Efficiency (εc), Current Density (CD) normalized on the internal volume of the insulated reactor (m^3), Energy (E), Chemical Oxygen Demand (COD) reduction, 0.14 < H$_2$ price < 0.40 €/L. E/m$_{Hydrogen\ (H2)}$ = 120 MJ/kg, E/m$_{CH4}$ = 50 MJ/kg, E/m$_{gasoline}$ = 44 MJ/kg, E/m = C$_2$H$_5$OH = 27 MJ/kg.

Neptune parabolic antennas would receive the complex sonic wave (IS/AF/US) of Juno 3 remnant of the explosions for releasing it onto Jupiter piezo-oscillator for which Magneto HydroDynamics (MHDs) would be suitable for converting chemical energy of an exploding bomb (e.g., Mars Nitro-glycerine [C$_3$H$_5$N$_3$O$_9$] = 227 g/mol, E = - 334 kJ/mol) into pulsed electrical energy E = 1 J/g (https://link.springer.com/article/10.1007/BF00800614). Surface antigens may be recognized by the antibodies into the cardiovascular system condensed around the thyroid, instantaneous pumping the heart instead of exploding shouting, as her just her would do looking after *'The two Olympic brothers'* (KDP Amazon 1.36 € https://www.amazon.com/dp/B0917SRFL2 and/or).

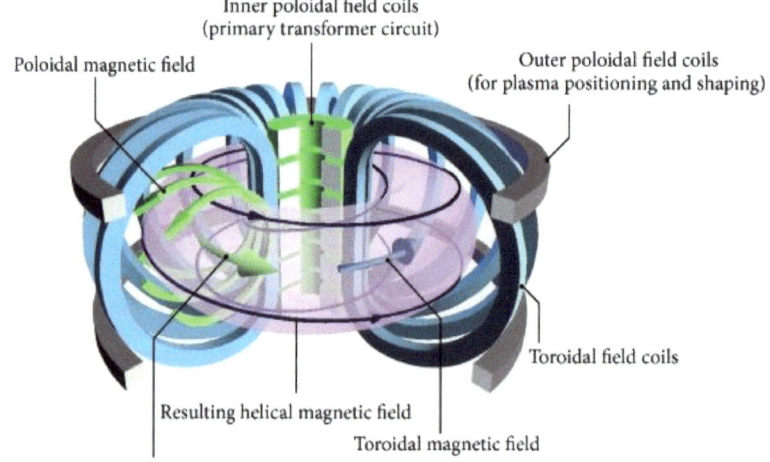

Figure 13 **Tokomak** confines ionised gaseous clouds converted via extreme heating into neutral plasma in toroidal (circular solenoid) magnetic field (B) stabilizing radioisotopes, acting as both as contamination prevention and radioactive waste recycling, made for producing thermonuclear fusion power. A helical magnetic field around the torus stabilizing the plasma current giving it a direction along the magnetic field of the circular solenoid. The optimal yield happened when the gaining fusion factor breakeven (Q = 1) is overcame, having a resulting energy greater than the ones utilized for stabilizing the plasma. Charged particles (H$^+$ and e$^-$) in the torus magnetic field experience Lorentz force (F = qE + qv x B) and drift spirally across its lines, losing with a variable degree plasma confinement. The stability issue of the orbiting particles is overcame bending the solenoid loops (*stellarator*) or injecting a guiding electric current that accompanies the plasma along its way (*z-pinch*).

Cosmic Wave Background (CWB)

The electronic engineering behind the origin of the Bubble Universe in which we are living and the hypothetic 'brain of the creation' in the bridge WormHole that let us be since 13.8 billion years ago at the Big Bang. **Cosmic Wave Background (CWB)** is the relic remnant frequency ν$_{CWB}$ ≈ 160 GHz the whole Bubble Universe dated in the recombination epoch for which intrinsically sealing the power of the alchemic change within it, which first peak determine the Minkowski spacetime curvature. The CWB bent the 3D network merging past and future in a sliding Singularity into the magnetic inductive toroidal WormHole, the same that in a reduced scale characterize your heart magnetic field. The Universal microwave frequency generated as opaque, hot and dense fissing plasma forming radiation ionizing hydrogen from a parallel Universe connected with us via an Einstein-Rosen bridge WormHole at time of the Big Bang, after which the Universe inflation led to the creation of transparent neutral hydrogen formation $H_{2(g)}$ gaseous clouds through which boson photon rays of *Helios* light diffused freely reducing from their highest energetic value to lower once keep redshifting, perhaps, totally vanishing the Universe in which we are living in.

Micro-oven, electric furnace that heat and cook foods homogeneously the exited outer surface (25 < h < 38 mm) via microwaves electromagnetic radiation due to, the viands. Its main components are: (Neptune transformer or converter), Saturn capacitor, Uranus LED, Jupiter metal stirring chamber similar to a Faraday cage cavity called magnetron, and digital control Aeolus fan, noise filter and ground wood (Percy Spencer – "Radarange", 1946). Microwave (300 MHz < ν < 300 GHz) non-ionizing radiation transmitted induce dipoles rotation exothermically in the cavity magnetron, that vary from maximum to minimum amplitude (A) and *vice versa*. Microwave heating is more efficient on $H_2O_{(l)}$ than $H_2O_{(s)}$ depending on the dipole moment [Debye]. Dielectrically heating of oils like the plasmalemma phospholipids and cytosolic water is accomplished by the microwaves like CWB. Saccharides and fatty acids absorb microwaves due to their HydrOxyl (OH) and Ester (ROR) groups, respectively. M-Oven can be used as Autoclave alternative for spore forming bacteria such as *Bacillus cereus* (P = 1 kW, t = 4 s) (J. Tachè, 2014).

Using safety precaution aluminium (Al) hat, glasses and alb and tools such as insulating tape, epoxy glue and the oven recycled metallic walls as waveguide it would be possible to convert the oven into a '**CWB H_2-generator**' that would simulate the radiation filling the Bubble Universe at the origin when coupled with the 'WormHole *in vitro*' (Kindle e-book 3.63 € https://www.amazon.com/Wormhole-vitro-Antonio-Silvestro-ebook/dp/B08DYDVZN9/ref=sr_1_5?dchild=1&qid=1596732888&refinements=p_27%3AAntonio+Silvestro&s=digital-text&sr=1-5&text=Antonio+Silvestro) itself mimic the once that generate the whole universe that we know and live within. The CWB should have been bending the special cylindrical proto-Einstein-Rosen bridge into the one that we all know heating it along its electromagnetic shielding walls separate the white with black Singularities, past and future in the silence of the present into which Zeus plasma filament discharge have been generated in a Ultra High Vacuum (UHV) leaving the formation of biological macromolecule like CH_4 and NO_3 and H_2, this last escaping where fission reaction let deuterium and tritium to react forming He noble medium extremely hot and dense gamma rays expanding in volume the Styropyrene foam compressing the insulated Proton Exchange Membrane (PEM) Nafion ($C_7HF_{13}O_5S \cdot C_2F_4$) cyclotronically displacing electron into the inductive holes, on contrary condensing its energy when relaxed into the primordial general configuration.

Structure of Magnetrons

ANTENNA — RF output

MOUNTING PLATE — Setting magnetron to oven

YOKE — Magnetic circuit

MAGNET — Generator of magnetic field

STEM — Input insulation and supporting filament

FILTER BOX — Shield of microwave leakage from stem

GASKET — Contact to wave-guide coupler of MW oven

FILAMENT — Source of thermal electron emission

RADIATOR — Heat sink

ANODE — Resonant cavity

TERMINAL — Input of anode and filament voltage

FILTER — Line conductive noise suppressor

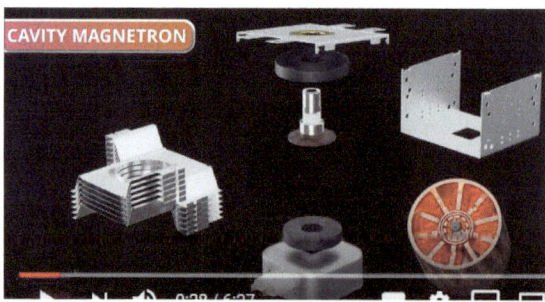

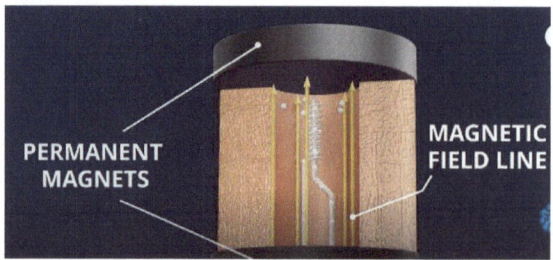

PERMANENT MAGNETS

MAGNETIC FIELD LINE

'Wormhole *in vitro*' © Copyright Antonio Silvestro, 2020

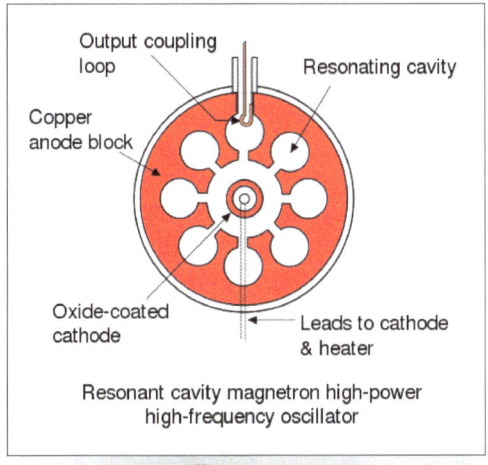

Figure 14 Jupiter magnetron vacuum valve tube characterized by a revolver circular deca-cavity with oxi-cathode (-) central pink ceramic Beryllium Oxide (BeO) potentially tumour ring substitutable with less toxic Aluminum Nitride (AlN) or dielectric insulator (Table 1 Kindle eBook 0.86 € *'Electronic circuits fundamentals'* https://www.amazon.com/Electronic-circuits-fundamentals-Antonio-Silvestro-ebook/dp/B086SFHDNV/ref=sr_1_10?dchild=1&qid=1596732888&refinements=p_27%3AAntonio+Silvestro&s=digital-text&sr=1-10&text=Antonio+Silvestro) and conductive anode (+) melted and solidified filling block on which surface travel the *Alternate Current* (AC) stored in the parallel heat sinking capacitative plates and induced into the cylinder in a resonant LC circuit (v_S = 3 GHz, P ≤ 25 kW) enclosed between two permanent magnets (NdFeB) - Image Source: https://www.youtube.com/watch?v=bUsS5KUMLvw

'Wormhole *in vitro*' © Copyright Antonio Silvestro, 2020

'Wormhole *in vitro*' © Copyright Antonio Silvestro, 2020

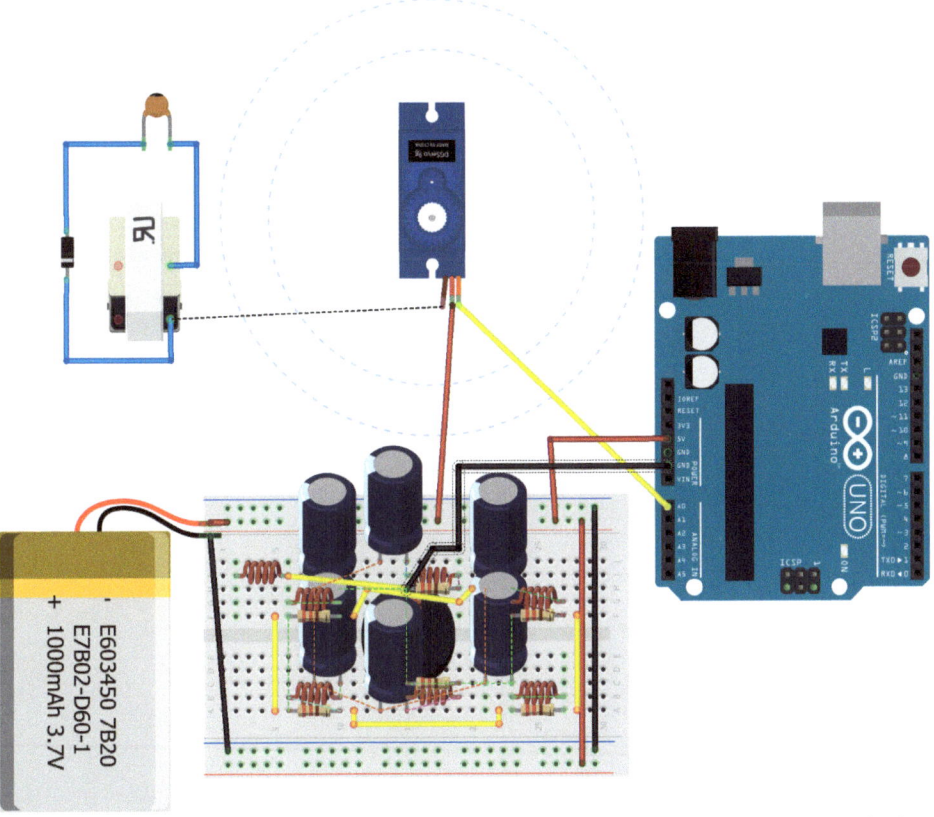

Figure 15 'Cosmic Wave background (CWB) generator' characterized by Jupiter a servo motor shaft spinning into the Juno hexagonal shield when the resistive Eurynome (R), the inductive Demeter (L), and the capacitative Leto (C) make together the RLC electromagnetic resonant e-circuit, otherwise, seen as two Wye Y triple phase transformers, substituting the anodic block of the cavity magnetron, from which Zeus lightenings arcs at frequency in the order of the kHz between the electrodes gap defining a double helix under the rotation of the motor wirelessly activated by an Arduino UNO, 9 V battery supplied, representing the brain in the far past of the White Hole in entanglement with the one of the Einstein-Rosen bridge of the Big Bang.
The double spiral from the Neodymium (NdFB) magnetic Singularity that sparked life into the Bubble Universe would have been accomplished a bi-path: one cyclotronic radiation along the normal relative jet by the *Hera* inductive (L_7) negative cathode (-), insulated with a metal outer ring preventing beryllioses touching it, and another along the output microwave RF inductive loop antenna walls of the Dirac-Einstein cone (or hexagon based pyramid imaginary bluish-violetish waveguide visible just when the motor is actively rotating the real electronic circuit in a complex CWB wavefunction generation - not shown in the picture) by the positive anode (+), with electrical arcs defining the accretion disk of the Black Hole of the Universe in which life is, the same characterizing your Nirvanic subtle body of rebirth.

SuperNovas (SN), Milky Way (MW) and Dark matter

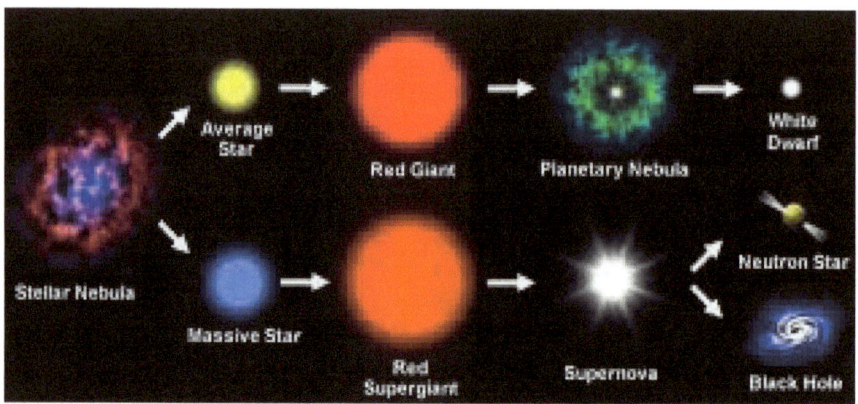

Figure 16 Life cycle of a star.

Stars can fall down at velocity of v = 7 · 10^4 km/s increasing their temperature and density, the elastic shells are progressively ejected out leaving SuperNovas, Neutron Stars or Black Holes depending on their initial mass: e⁻ O-Ne-Mg collapse and neutron stars' remnants 10 < Progenitor star M$_\odot$ < 250 photo-disintegration collapse and massive black hole remnants.

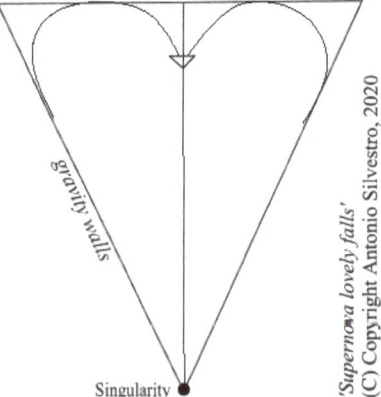

Figure 17 *'Supernova lovely falls'* – Einstein/Dirac cone with gravity pressing onto the event horizon - Image source: © Copyright Antonio Silvestro, 2020.

SuperNovas (SN) named by Walter Baade and Fritz Zwicky of the Californian Institute of Technology (Caltech) in 1931, are transients powerful, luminous, massive, giant, degenerating stars in which generative re-ignition runaway nuclear fusion let them exothermically explode, radially ejecting the old outer-layer several times faster than light, collapsing into its own, leaving just a young neutron stars' core, that bends the gravitational oscillating field falling in the Dirac light cone of the past, curving the space-time stabilizing itself in the present surface as singularity of a supermassive black holes that rotating around its polar z-axis, would accrete the remnant gaseous matter and its emitted cosmic radiation around its void, empty of any charge, actively attracting them along a shocking conical spiral, becoming the galaxy-generating quasar [*Quasi-Stellar Object* (QSO)] like the once that 13.6 BYA gave origin to the *Milky Way* (MW), the about 250 billion punctual stars placed helically enlightening the cone of the future: the white hole.

Super massive *neutrons stars* reach T = 10^6 °C on its surface, done of atmosphere, Fe-Gnocchi-Spaghetti-Lasagna phases crust (nuclear pasta), quark-gluon plasma and strange quark core. Their spin induces pulsing beam of RF ('Magnetar' B = 10^{10} T), rotating among themselves, crushing into *'Kilonova'* leading the nuclear fusion making heavy nuclei like ^{79}Au, ^{92}Ur, ^{78}Pt, ^{67}Ho, ^{83}Bi, ^{77}Ir collapsing into a Black Hole.

SNs occur in the Milky Way on average of three times per century. One of the five telescopically observed SuperNovas anterior to the eye-naked visible for 185 days Cassiopeia SN 1181 (galactic coordinate g.130.719+03.084), SN 185 Centaurus, SN 386 Sagittarius, SN 393 Scorpius, SN 106 Lupus, SN 1054 Taurus, should have been the SN associated with the formation of the **'Tr-J Tezcatlipoca Black Hole'** decoded on the Aztec glyphs of the Sun stone almanac acting as primordial telescope (for more info about it *'Neo Wave necklace – Crimson aurora'* (bead I) © Copyright Antonio Silvestro, 2019 https://www.amazon.com/s?k=antonio+silvestro&i=digital-text&ref=nb_sb_noss), among which the author would guess between Scorpius and Taurus, leaving to this last the most plausible, but certainly, it could investigate deeper with the aforementioned detecting devices whenever economical support would be enough.

A binary stair pair orbiting along the same plane, grounded on the same barycentre, in opposite direction, till one absorbs the atmosphere fragment of the other into itself, becoming relatively *red giant*, this great amount of matter and radiation is strep-out, accreted, firming the orbit leaving the two stars well bordered all around, the giant collapse onto itself becoming a small, but bright *white dwarf* ($M_\odot$ = 1.44), C-O shelled, O-Ne-Mg cored, this encircled by the companion undergoes **runaway nuclear fusion**, explodes releasing (about $10^{43} < E_w < 7.5 \cdot 10^{44}$ J), ejecting it away, shocking the Universe with a slow, but massive wave of velocity v = 12.500 km/s and absolute magnitude - 19 < M < - 15 which luminosity is caused by the *radioactive decay* of Ni -> Co -> Fe emitting gamma rays photons inducing the oxidation of nitrogen in the thermosphere which *Nitric Oxide* (NO) so produced precipitating is later entrapped in the polar ice and deplete the ozone, it let invade UV-B through the stratosphere to the beings living on the Earth planet polluting them with its tachycardic effect. The *type of SuperNova* influences the kind of elements released in the surrounding Universe (e.g., Ia: Fe, Si and Ni, core collapse: O, Ne, Ga, II: H, He). Stars hot plasma collapse in the core fusing H into He, medium stars like the Sun (T = $6 \cdot 10^3$ °C), burn He into C and O_2 becoming red giant expanding their volume, eventually, turning into smaller white dwarfs. *Burning series* from the outer layer to the star core: C, Ne, O, Si and Fe, that should be related respectively to the planet Mercury, Neptune, Uranus, Jupiter and Mars.

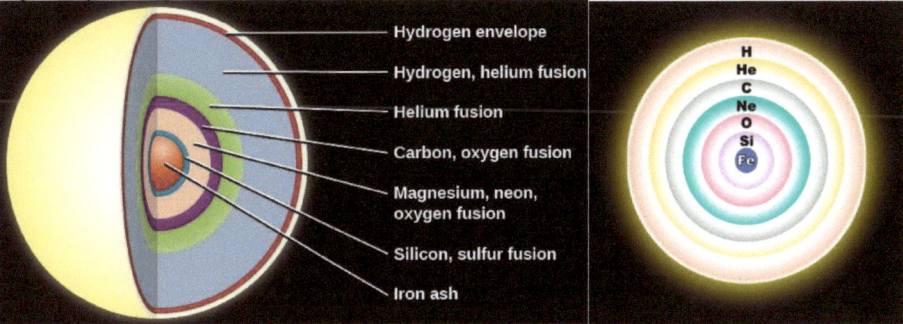

Figure 18 Massive Star layers - Image source: https://courses.lumenlearning.com/ and Rursur R.J. Hall CC BY SA 2.5 https://en.wikipedia.org/

'Wormhole in vitro' © Copyright Antonio Silvestro, 2020

Milky Way in the Babylonian epic poem Enûma Eliš come from the slayed Dragoness Tiamat, while, according the Greek mythology is spilled out the breast of *Hera*, sister and wife of the Greek God *Zeus*, when the trickster *Hermes* let suck *Heracles* the boobs of the Goddess that freeing herself from the son of the husband token with *Alcmene* flowed in the Universe (MW - Greek: γαλαξίας κύκλος, Latin: Via Lactea) (solar mass$_{MW}$ = 10^{12} M$_\odot$ = 2 × 10^{42} kg), scientifically is a hazy white central-barred spiral galaxy distant from Earth 15 · 10^4 < ø < 30 · 10^4 ly (1 ly ≅ 10^{13} km) and visible from itself if the sky is darker than 20 M/arcseconds, done of ≈ 250 · 10^9 stars (5 · 10^{10} M$_\odot$), separated by interstellar gas (90 % H and 10 % He) weighting 10 % of their mass, and ≥ 100 · 10^9 planets, rotating together at 600 km/s, in which the slower *Solar System* (v_{SS} = 220 km/s) is placed, precisely, in the Orio-Cygnus Arm, 27 · 10^3 ly to the *Galactic Centre* (GC) Sagittarius A, a Super Massive Black Hole (solar mass$_{SMBH}$ = 4 · 10^9 M$_\odot$) emitting Radio Frequencies 20 kHz < v_{RF} < 300 GHz, accreted as an inactive galactic nucleus with a ratio of 10^{-5} M$_\odot$/year, with the Sun ≈ 57 ly Norther than the GC harmonically oscillating ≈ 2.7 times per elliptical orbit around the *Galactic Plane* (GP), each lasting 240 · 10^9 years. Elliptic function: $f(z) = f(z + \omega_1) = f(z + \omega_2)$ ∀ z ∈ ℂ, where: ω_1 and ω_2 = non-zero complex numbers and $\omega_1/\omega_2 \notin \mathbb{R}$.

Conjectural cloudy gaseous astronomical **Dark Matter** (1.5 · 10^{11} < solar masses < 4.5 M$_\odot$ · 10^{12}) do not follow the Keplerian dynamics being in constant rotation at speed equal to the solar system v_{DM} = v_{SS} = 220 km/s account 90 % of the whole MW. Old stars form globular clusters orbiting with the same angular velocity relative to the common barycentre as set of astronomical punctual bodies of an equilateral polygon in a *Klemperer rosette* configuration form the stellar halo of the MW (1 < T < 2.5 10^9 K, - 17 < T < - 15 · 10^9 °C) like the tetra-loop of the *Orion constellation* through the resonating human vertebra of the adorer. Nucleon cosmos-chronology permit the *MW dating* via the measurement of radioactive elements like Thorium (^{232}Th) and Uranium (^{238}U) 12.6 · 10^9 years ago.

Artificial wormhole

Einstein Field Equation (EFE) are 10 equations for which time seems to curve the space in which it has been defined by one or many energetic massive corpuscles. The EFE relate the inductive toroidal geometry of the Minkowski spacetime to the matter displacement depending on Big Bang elementary particles charge and mass:

$$G_{\mu\nu} + \Delta g_{\mu\nu} = (R_{\mu\nu} - \tfrac{1}{2} R g_{\mu\nu}) + \Delta g_{\mu\nu} = k T_{\mu\nu} = \frac{8\pi G}{c^4} T_{\mu\nu} = \frac{8\pi F_b \frac{r^2}{m_w m_b}}{c^4} T_{\mu\nu}$$

Where:
$G_{\mu\nu}$ = Einstein tensor curvature of pseudo Reimann manifold
$g_{\mu\nu}$ = metric tensor embracing all defined geometries
$T_{\mu\nu}$ = stress-energy tensor
Λ = cosmological constant representing the energy of the hyperspace vacuum
G = Newtonian-Cavendish universal gravitational constant = 6.67 · 10^{-8} L/kg s^2
c = speed of light = 3 · 10^8 km/s
r = distance between supermassive White and Black Hole singularities = double of Swarzchild radius of two observable Universes = $2r_s$ = 4 G(M$_\odot$/c^2) = 2 G (E) = 1.3 · 10^{26} m
F_w = White Hole force = F_b = Black Hole force
m_w = m_b ≈ mass of Jupiter = 1.8 · 10^{27} kg
g = gravity = 9.81 m · s^2

On the gravitational wave generated on the Einstein present hypersurface of the Big Bang due to the instantaneous wormhole formation have been possible to design any enclosed geometry and adapting the Pitagoric theorem to the tensor curvature, manifold continuous oscillating, perhaps, due to the solar wind and/or the Sun antipodal in the White Hole side of the Bubble Universe. But what there is in the other side, our most far past and the reverse of ourselves as the "red Indian" throwing the stone against gravity in the water let it undulate looking his own face stretched appearing weirdo as a "Yankee". The elastic membrane would have been inhaled and exhaled along the vital energy of the sentient being living from one and another side of the Universe sealed in a common Bubble as the two first cells originated from the division of the zygote surrounded by a common plasmalemma. Life, here as there, a complex wave of events, positive and negative that make you singular. The Singularity (S) bounce on the membrane and jump to the other side via a hyperbolic path inside the torus magnetic field with centre is dated 13.8 GYA, but it has to have been characterized by a semipermeable membrane through which RNA retroviruses and anaerobe autotrophs would have been passed through recombining over the fringe from the White to the Black side, and vice versa, for which the origin of LUCA cannot be surely referred to the side in which we live, the reducing cathodic one (+). Nevertheless, that the spacetime curvature that hide the moon behind the Sun as its empirical significance just on planetary scale, reason for which despite the EFE may be applied to any 3D physical object on a plane, approximate the real surface as a flat plane is the conceptual solution of the undulated complex Bubble Universe in which we humans leave into, totally pragmatically useless for virologist resolving nanometric distances and for ordinary life itself.

The collapse of the iron (Fe) core of a star induces its implosion in a fraction of seconds dying in as supernova release its fragments remnants into the cosmos leaving its neutron core (1.5 < Solar Mass < 3 $M_\odot$) becoming massive when the outbreaks gravitational kickback falls onto it, rebirthing as neutron star of a relative heavier **Black Holes**, fluid ellipsoidal mirror with strong gravitational field giving the direction to the diverging flow of among the dimension of two mixing physical Universe in entanglement that let displace through it discrete masseuse and massless sub-atomic particles and the weaving radiations deriving from them around the accretion disk heated by the turning friction. Perhaps, the punctual entity with the strongest capability to symmetrically bent the gravitational field is the Singularity of a Black Hole astro-physical phenomena for which energy and related physical quantities can be calculated at the intersection of the gravitational strings using tensor equations based on mass and its movement in the empty space where the inertia reaches its minimum values in an ideal never ending Newtonian continuum accelerated perturbative motion along the natural geodesic curves laying onto Riemannian topologically closed 3D surfaces, otherwise, manifolds. Thermonuclear reactions, fission and fusion happen within the Black Holes and with high probability all types of decay occur, according which star is intake, seldom releasing the excess energy, among which quantum of light – 'boson photon', in coherence with the Albert Einstein 'Theory of General Relativity' that difficult the motion as plasma solar corona jets-like. At their origin, this sucking astronomical objects, are characterized by an intrinsic power $P = E/t = m\ c^2/t = 19 \cdot 10^{26}$ W => $m = E/c^2$, corresponding to a star with a diameter $\emptyset \geq 500 \cdot 10^6$ km, in agreement with John Michelle observation (https://youtu.be/e-P5IFTqB98). The specularity of the Black Holes is highlighted with their distortive passage through the border of a galaxy that resolute the switch between the entering and exiting prolate ellipsoid having the event horizon as symmetry axis, actually, the pusher of the galactic confine. In other words, the Black Holes inseminate the invading galaxy with the radiation captured in the one where it came from. The rotatory displacement of this astronomical entities continuing straight when their energy storage tank is filled blowing apart the useless bright ionized gaseous clouds with a relatively lower escape velocity, that starts orbiting around it, being alone less able to challenge the gravitational field of the galaxy. These astronomical dark objects are not just dominating the empty space, but in mutual symbiosis with it exchanging along the semi-permeable membranous Event Horizon real particles for virtual ones. In other words, just the virtual images of an empty space particle can move out of the Black Holes according its independent behaviour of annihilation as real corpuscles dissipating the Black Hole energy; but, certainly, for voluminous and supermassive Black Holes this **Hawking emissive phenomenon** of the quantum field theory in curved Einsteinian space-time would correspond more to a filtering process then an energy loss phenomenon. The **rotating Kerr and Kerr-Newmann black holes** (J > 0) would have been characterized by a merging *vescica pieces* enveloped by a thought ergosphere; their spinning can be described by mass-energy (M), linear (P) and angular momentum (J), position (X) and electric charge (Q). Two black holes rotating counter-clockwise inscribed in an ellipsoidal shell fusing themselves form supermassive Black Holes of million Solar Masses (1 $M_\odot$ = 2 · 10^{30} kg), giving origin to galaxies. In this last, the plus (BH^+) and minus (BH^-) primordial Black Holes conjugated, knotted together in a chiasmatic singularity at the cone vertex, are confined in the galactic gravitational field from which they try to escape in opposite directions on the axis perpendicular to the that intersect the singularity, bending the galactic gravity walls with their prolate orbital lenses ('p-lenses'). Being the BH gravity field relatively low compared with the galactic field, everything in them stretches, increasing its volume make the Black Holes chaotically increase in their entropy in a stabilizing dynamic equilibrium with the break-down of stars integrated. The black body thermal radiation surrounding a spinning black hole emitting electromagnetic radiation for the accession disk, would interfere with it destructively, diffusing involute incandescent sparking breaking them down separately into red, orange, yellow individual wavelenght and in combination with the whole white spectra from its Event Horizon, during a variable time range according the space occupied, before sucking it inward to the originary basal Singularity (S) surrounded by a deep bluish-violet halo were the annihilation of space and time is better manifest.

According Gustav Kirchhoff a **black body emits radiation** in all the EM spectra, within this the peak of emission is equal to the Wien's displacement constant (≈ Star λ_{max} : T = 1 nm : 10 K):

$$b = \lambda_{max} T = \lambda_{max} \frac{L}{\sqrt{4 \pi R^2 \sigma}} = 2.9 \cdot 10^{-3} \, m \cdot K$$

Where:
L = luminosity (e.g., L_{Sun} = 3.83 · 10^{26} W)
R = star radius [m]
σ = Stefan – Boltzmann constant = 5.7 · 10^{-8} W/m²/K⁴

The fluidodynamics of a *flame* shows how it shapes according the minimal friction, where the drag to its motion is the smallest possible and the fact that the breaking down of the originary energetic occurred with lowest frequency at the antipode of the igniting event beigns it having been already converted into radiation along the main axis of the imaginary ellipses bordering the real flame in its complex manifestation. Matter-Radiation transition happen according the wave-particle duality that merge General Relativity with Quantum mechanics. A flame would burn the air in a positive feedback triggered by oxygen, while, the exact opposite happens with the Singularity cooling the vacuum incrementing the depression triggered by its gravitational field. As particles tent to escape the aspiring evolute circle along involute curves, so the Black Hole leave impure remnant out from its Singularity, obligate to tumble themselves far away, nuclei rejected by the supermassive for becoming, perhaps, the source of new galaxies, activating themselves in another distant favourable astronomical environment. Bluish electric sparks would have been diffused along the evolvent in the insulating extreme vacuum of the generative Bubble Universe shaping the bluish sphere of the Singularity of infinite gravitational field submerged in the zero-G, when applying an electric field that exceeds the dielectric breakdown strength of the gap. Actually, the involute curve would be just the average linear free mean path route that the corpuscle that instead accomplish a periodic course through it. Free electrons of Cosmic Background Radiation and Rays flows continuously describing an half Jacob elliptic curve between the positive and negative conductive poles of the Bubble

Universe as free e⁻ flows through the Earth atmosphere from the clouds to the ground, an immense potential, during the lightening (I = 30 kA, q = 15 C, E = 1 GJ), in kitchen or internal combustion spark plugs, or stun gun (ΔV = 150 kV, air gap = 2.5 cm). *Cyborgs* with capability of emitting electricity from the eyes could be designed basing themselves onto the aforementioned phenomena of electric sparks [average adult *Pupillary Distance* (PD) = 6.4 cm] or even higher voltage ionizing glow discharge.

Remember when you were young?
You shone like the sun
Shine on, you crazy diamond
Now there's a look in your eyes
Like black holes in the sky
'Shine on, you crazy diamond by Pink Floyd'

Figure 19 Herr. K. Schwarzschild and the 2D view of the Black Hole. Schwarzschild radius $r_s = 2G(M_\odot/c^2) = 2$ G (E) ≈ 2.95 M/M$_{Sun}$ km ≤ prolate orbital ellipsoid lenses radius (Black Hole radius) $r_p = r$, where: G = Henry Cavendish Gravitational constant = 6.7 10⁻¹¹ m³/kg s², c = Albert Einstein speed of light = 10⁸ m/s (e.g., the *Nirvana Sharir* of an average *Homo atm* and even inferior *Homo sapiens* would have been characterized by a Schwarzschild of 10⁻²⁵ m).

Gravitational singularity can be shaped differently, acquiring for example conical finite singularity over the differences in shape where the rough spacetime find itself such as cosmic strings complexly connecting manifold in the void and the Schwarzschild Black Hole or curvature like the ring Kerr singularity, that naked of event horizon, hence, neither evaporating nor decay of Hawking radiation last forever, expressed in Boyer-Lindquist or Reisner-Nordstrom metrical coordinates, a spinning Black Holes in the void that can theoretically become a Wormhole. The time t = 0 of the Bing Bang cosmological model was characterized by a dimensionless magnetic monopole singularity, but infinite density and temperature bending without limits the gravitational field, acquiring one dimension along one gravity wall of the light cone for later seal its domain in the second dimension marking its 3D empty volume. Its **radius** would be equal to:

$$r_S = \mu \pm (\mu^2 - a^2)^{1/2} = \frac{GM}{c^2} \pm (\frac{GM}{c^2} - \frac{J}{Mc})$$

Where:
μ = GM/ c²
J = angular momentum

$a = J/Mc = Q\sqrt{G}/(4\pi\epsilon_0 c_4)$

The probability for the singularity to escape a hole reaching the opposite charged through the wormhole instantaneously, triggered by the heterodyning process of wave combination is shown in its initial logarithmic phases cryogenic, when the surrounding primordial Universe medium would inflate lowering its temperature for firmly fix it in place. The natural fundamental constant in period (T) oscillation of the Singularity would be in an instant rectified (AC->DC) according the **Laplace transform**, moment during which the fringe between the past and future would be accomplished permitting theoretically time-travel also back in time in a unitary step corresponding to the diameter of the Dirac cone, otherwise the **double of the Schwarzschild radius**.

$$L(A) = 2\, r_s : \int_0^\infty f(t)\, e^{-At}\, dt = \frac{GM}{c^2}$$

When the step would be one [L (A)], the wormhole would fold on itself in one only cone in which conformation the fall would be as the one show in the 'SuperNova fall' to the vertex for later rototraslating capillary along the walls for emerging again, in future or past. Have you ever practiced *Paschimottanasana* or *Uttanasana*, Well, if so, you may now how your halo envelope your Sthula Sharir touching with forehead on the knee gently falling? Anyway, the transition, that would, perhaps, one day soon teleport objects like your money and maybe even sentient organism through time and space in the Minkowski spacetime as Einstein quantum of light. So, do not be heavy and believe, this the first condition for continuously rebirth in the immortality of scattered light through your subtle bodies. Could you feel the magnetic field of your babies far field even if silent? Could you touch the skin of your husband, could you see their eyes crying when you are not. Well, believe, it happens, they do miss you.

The wormhole breaths widening and reducing its volume within the general Minkowski spacetime, merely, let flow the Singularity between one hole to another when is relaxed, relaxation due to the particle composition of the medium as atropine does with lungs so the Einstein-Rosen bridge is freezed when isotopic transitioning Bequerellian decay happen Hydrogen->Deuterium->Tritium->3-Helium, finally, into 4-Helium superfluid cryocooling (violet sphere in the Figure 25).

Non-Relativistic Quantum Mechanics (N-RQM) applicable in *'Galilean invariance relativity'* where all the laws of motion are in a common inertial frame, while, RQM works in *'Einstein special relativity'* where gravity curve the space-time. Since that, mass and weight are the reference physical quantity, respectively. Their combination in the **Quantum Field Theory (QFT)** become suitable for quasi-particles and subatomic particles. Vector Euclidean analysis, tensor absolute differential Ricci calculus and Fourier transformation are needed for representing radiation functions.

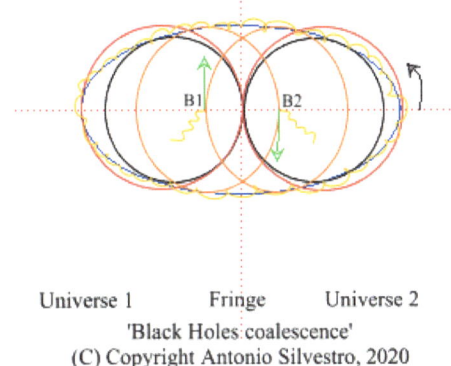

Universe 1　　　Fringe　　　Universe 2
'Black Holes coalescence'
(C) Copyright Antonio Silvestro, 2020

Figure 20 *'Black Holes coalescence'* – Images Source: © Copyright Antonio Silvestro, 2020.

The corner between two wave functions (Ψ^+ and Ψ^-) like the two sinusoidal strings of the **DeoxyriboNucleic Acid (DNA)** is defined by the inner products, a positive real number as devised by Niels Bohr and Werner Heisenberg in the **'Copenhagen interpretation'**, considering the total charge of the two phosphate backbone sequences depending on the opposite polarity of the nitrogenised bases purine-pyrimidine pairing among the two complementary filaments, precisely, equal to:

$$\Psi^+ \times \Psi^- = |\Psi|^2 = P(\Psi^+ \to \Psi^-)$$

$$\Psi = \sum_n a_n \Psi_n$$

Where:
Ψ_n = orthogonal wave (e.g., DNA positive strand)
a_n = orthogonal coefficient = $(\Psi_n, \Psi)/(\Psi_n, \Psi_n)$

Black Hole whirling coalescence can be represented mathematically as the 'Ψ_2 collapsing' to a new singularity Ψ_1, arisen from rooting characteristic values scalars set wave, hence, with no-zero solution (e.g., root note of a scale), as the two entwined *CoDing Sequences* (CDS) of the DNA, without unbalancing mutative loops, veil a unique stable protein. Would be precisely the DNA into the nucleus of the cells of the lymphatic system though all over the Nirvanic subtle body to silencing or expressing quantum superposition among the strands of opposite charge. The amino acid sequence expressed ($\sum_{-C}^{-N}$ **aa**) [where: -N = -NH$_2$ = amine terminus and –C = -CO$_2$H = carboxylic terminus and the Black Hole radius (r)] can be described by the following **quantum state abstract vector** in Dirac Bra-ket notation, passing through solenoids, Pauling-Corey-Branson α-helixes and the Einstein temporal relativistic cones, respectively:

$$|\Psi(t)\rangle = \int dx\, \Psi(x,t)|x\rangle$$

Set of **momentum**, mass per velocity ($p = m \cdot v = -i\hbar\nabla$) of the wavefunctions of the displacing Black Holes one against another is infinite as the equation clarify:

$$\{\Psi_p(x,t), -\infty \leq p \leq +\infty\}$$

Quantum vacuum state fluctuation are the temporary changes carried by the four fundamental forces (electromagnetic, weak, strong and gravity), defined by Werner Heisenberg-uncertainty principle that chaotically took place when the Bubble Universe inflate itself during the beginning of the Big Bang (1.6 GYA), highlighted by hydrogen (H_2) energy states Willis Lamb shift. These virtual particles exchanged for real one at the Black and White Hole horizons violate the conservation of energy principles has they are placed in the Einstein-Rosen bridge of both creation and destruction at same time as the Schrödinger cat paradox would exemplify. Instantaneous changes of the whole as the null would have been manifested continuous corpuscular annihilation making impossible to determine momentum and position in the same instant, if not fixing this last. For which, cryogenizing the time ov the originary universal event in the empty space (t_o), maybe, it should be possible overcoming the *Heisenberg uncertainty principle* calculating contemporarily both position O (x, y, z) and momentum$_O$ of a lepton electron (e⁻) in an orbital p, fixing the position to the intersection between the ellipse and the main axis. O (x, y, z) corpuscle laying on the intersection point between the ellipse perimeter and the secondary axis of the ellipse itself characterized by a momentum:

$$M_O = m \cdot v = m \cdot (f_{xy}/t) = m \cdot [-(p/2)/t] = m \cdot -2\pi t \sqrt{a^2 + b^2} \text{ [kg s m]}$$

Where:
m = mass [kg]
v = velocity [m/s]
f_{xy} = ellipsoidal loop length [m]
t = time [s]
p = ellipse perimeter $\cong 2\pi \sqrt{a^2 + b^2}/2$
a = longer axis [m]
b = shorter axis [m]

Jean-Baptiste le Rond D'Alembert in the XVIII century defined an equation for determining the way of changing of a linear mono-dimensional wave resolved both in its additive sum and multiples *superposition states* like the **'Black Hole whirling coalescence'**. The meeting of two black holes could let the interaction of the related wormholes merging two different Universes into one a catastrophic interaction that would let them spin coherently in a domain between the fringe where the Singularity is. Would be it be teleported? Well, that's to be answered. But what is sure is that it was hot enough for being melted in a pizza hoven, heavy as radio-fluorescent Uranium, and sparking as a Nikola Tesla Globe.

A discontinuity point in the function of the **wormhole**, right in the middle of the Black/White hole border is the singularity, the Doppler red shifted where light cannot be seen any more being gravitationally entrapped. Precisely, the singularity would be an inflection point where the function change orientation, overcoming the past, in the present instant, to the future. Astronomical objects and their radiations falling into the Black Hole along a wave that increase its length approaching the bottom singularity, this the origin of new galaxies and celestial objects emerging through the white hole. Hyperbolic flow along the lines of a toroidal magnetic field due to the motion of the ElectroMagnetic objects displacing through the wormhole according the Faraday law of inductions.

The top and the bottom of the clepsydrae Wormhole would have been characterized by **lenses** placed into the accretion disk perpendicularly to the pH-timeline. The White Hole, would be merely a liquid water mirror, while, the Black Hole a gaseous hydrogen once. Flat, concave, convex, bi-sided or mono-sided lenses depending on the polarization direction yearned, but the most congruent with the primordial reality in an oscillatory probabilistic dynamic state changeable among them as the waves of the oceans where Poseidon rules, leaving the concave lenses the once that would seal in it all the possible cases suitable for overall theoretical calculi, but the mono-concaves concealing also the upside-down motion of the clepsydra.

Figure 21 Concave lens placed onto a transparent plastic real cone (or truncated cylinder), for empirically highlighting the complexity of the Minkowski spacetime curvature and the Einstein-Rosen bridge just looking through it noticing that the smaller diameter is magnified appearing in a wider reflected cylindrical image instead of a real cone.

The inside of the Singularity would be reddish, while, out of it bluish due to Willis–Tyndall **scattering** for which the elementary particles in a colloid's suspension, blue is emitted and absorbed much more strongly than the red, furthermore, similarly, Rayleigh light scattering the intensity of the electromagnetic radiation of their shah would depend on the fourth power of the frequency coherently with the inverse law of the radius, hence, the distance from the center of the Singularity. The blue monochromatic radiation carrying the genome of the LUCA reference descendant from the generative spheroids in the far past to the future of your reading would be the once capable to teleport its quantum information superimposed between the two halves of the Einstein-Rosen bridge under the CWB generated by the undulation of the feeling hypersurface of the present artificially made by CWB generator in an LC circuit between the inductive photoelectrical cathode lenses (+) and the capacitor done of the anode (-) and the plane wave shared by both cylinder-conical cells in the Minkowski spacetime. The middle-insulated undulating wave is generated by the stretching JJ effect due to the the pressure variation in the UHV Bubble Universe leading to displacement when hot and depressed as the Singularity, *similia similibus curantur* would say the ancient Latins healing the ignorance about the origin of the whole existence. The future *Archean, Eubacteria* and *Eukaria* new evolving or in evolution species like the *Homo atm* is continuous as all the light spectra respect to LUCA, while, discontinues as mono-chromatic light from it to themselves, highlighting that the step forward are flash light in the dark of the past. Hence, just with the whole spectrum we may find our common ancestors in the still not evolved future descendant teleportable in one only instantaneous Einstein-Dirac absolute unitary pulse. Genome traits expression/silencing duality due to light would have been the most diffuse mutation inducer phenomena through all over the Bubble

Universe evolution in the infinite gravity Singularity as in the UHV enveloping it.

The meeting among two Singularity would happen in a wormhole, when the resonating centre would bend the spacetime in a common field of forces merging together in the projection of the fourth special dimension where wishes flows over distances and kisses may be felt as here on your wet dry lips while thirsty of love. Via the Einstein-Rosen bridges you can meet the purity of your heart on the string of Jupiter, in the resonating *Manas Sharir* you can travel far field Fraunhofer distances in an instant that any quartz crystal in entanglement could mimic shielding any interferences. The plane surface of the planet Earth, could be bent and let pass the thinking via temporary wormhole in a continuum spacetime that just the capacitative soul can divide. Quasi-permanent intra-universe wormhole so defined by Sir. Hermann Minkowky, would be what many lovers on heart feeling each other could live. So, an astronomical entity, become a fleshy reality of a complex life phenomena. Thoughts would act as exotic matter travels in the wormhole according the Raychaudhuri's theorem revised for telepathy. Meeting that can happen just when the Atman of the partner are synchronized on the string of Saturn and the metric of the transverse empty tunnel follow the Ellis one. But that's not all, you may also see with closing eyes, when one of the two related tries to escape the heterodyning relationship as quantum of light are emitted by electrons exited in our midbrain assessable with Single-Photon Detector (SPD) which Quantum Efficiency = electrons/photons is positive (QE > 1), the Cartesian Gate that finally, split as in two individual as a wormhole, in black hosting and white pushing out energy-matter. Perhaps, the white hole would be what merge two black hole Nirvanic bodies of two individuals. Ying and yang, the man and the women merging in an imaginary and real singularity that instantaneously teleport itself from the white to the black hole, let appear the complex Dirac cone as filled by two separated singularity. Time travel could happen in a pure, uncontaminated singularity cell with semi-permeable walls that would change the potential along them varying the charge, hence, the direction of the cosmical solute within it during the jump across the Event Horizon along the axis passing through the II<->IV **Kruskal-Szekeres coordinates**. As magic combinatory game the two spheres would be at same time in two diverse places swapping each other instantaneously as thought do during the presence-absence of the mind, between two lovers. Controlling Chitta, silencing it from temporal interference the common space would manifest itself even for more than an instant, prolong in an iced wormhole which spatial conditions can perpetuate themselves for endless time. The relativistic spacetime concept would lay on this probability of meeting of the two singularity as one only through a theoretical continuous scattering bridge. One with the medium absorbing and emitting selectively real for virtual particles, and vice-versa, instantaneously changing voltage difference along the surrounding membrane during the instant of the time-travel. Dividing this instant into differential Newtonian-Leibnitz differential steps, each with an opposite charge, the oscillating envelope would be unrolled and the half-cone has been done of pass to the antipodal side destroying the singularity within it and recreating it to the other half, from *Sthula* to *Nirvana* in the humans with knowledge about the presence of the singularity in their chakras.

'Wormhole *in vitro*' © Copyright Antonio Silvestro, 2020

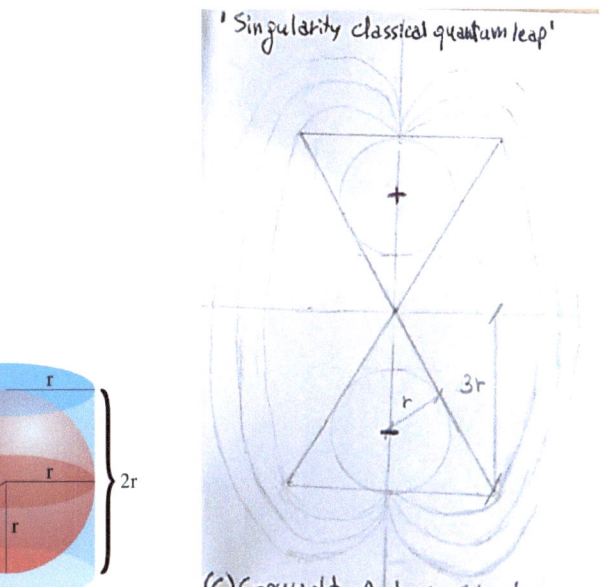

Figure 22 A sphere occupying 2/3 the volume and surface area of its circumscribing cylinder including its bases (left). Let fall into a bi-Dirac cone the singularity it wouldn't be blocked at the vertex, but bouncing to the otherside recirculating via the Newtonian apple magnetic field lines surrounding the wormhole, sealed into them, constricted levitating into the other half according the Archimedes' hydrodynamic law, but in the empty space. So, the instantaneous quantum leap of the singularity could be approximate by classical Renaissance physic too. Two as one, lonely, but paired in an everlasting entanglement that would let see them divided, just when frosted by the absolute observing it from the other side of the mirror (right) – Image source: © Copyright Antonio Silvestro, 2020.

The regulation of the size of the observer onto the hypersurface of the present would be essential for determining the amount of the charge deposited acting as in a salt bridge of an electrolytic cell widening while dispersed onto the hyperboloids of the light cone. Vary the angle of the cones of hyperbolae the electron transfer could be more or less fast, reaching its optimal simultaneity at $\alpha = 90°$ corresponding to and fixed unmoving event O ($v = 0$ m/s). Along the bisects of the cone, light travel along the $\alpha = 45°$ slope at speed of 1 light-second/s = $v = 3 \cdot 10^8$ m/s; precisely, light displace itself along the bisects with variable corners respect the z-axis, depending on its energy content: the low frequency event O with $\sigma = 90°$ roto-traslating on circular orbits, then $\varepsilon = 30°$ on an elliptical orbit, $\pi = 20°$ on parabolic orbit, $\eta = 10°$ on hyperbolic closed path confined in the walls of the cone previously described. You can imagine a quantum of light placed in the apex of the cone having the lowest energy value being static, suddenly spinning, accelerating along the polar axis, hence, in accelerating motion along the imaginary walls of the cone, becoming real with each circular loops; the tiny corpuscle, become flatter and flatter with each turn leaving its red ink placed onto the complex cone. Suddenly, its fall, just an instant, but fatal for its scheduled circular orbital plan, that see itself widening, expanding becoming elliptical, in need of more energy, but rewarding of more volume. At that time, creation was ongoing, orange die was diffusing along the walls of the cone.

'Wormhole *in vitro*' © Copyright Antonio Silvestro, 2020

The planets, stabilizing their natural and stable orbit in the system that we now currently like the Solar one. But light spread ever very in a continuum diffusion in a peaceful Universe, till scattering happen and the brightness of the white light become obscured by the relativity of the masseuse particles. The yellow range of the boson is imprinting the Einstein cone, highlighting an elliptical roto-traslatory pathway. When it falls again, raising along the bisectors in a quicker way, parabolically orbiting diffusing green, then cyan light till falling again for the last spreading of highest energy visible level in the indigo and violet hyperbolically. In each dimensions progress along a wave function the system cannot be rigid and linear but fluid and bent, but certainly, it seems easy to represent the Universe with the flat Euclidean space instead of a Hilbert one, considering lesser variables and of firm trends, but it isn't on the long run, because errors due to approximations accumulate themselves and what is going to be described is not more the curve spacetime complex Multi-Universe in which imaging it we are really living within it. In other words, to define objectively reality is needed to redefine every physical quantity on relativity diagrams in which the 3D axis is curved of hyperbolic angles.

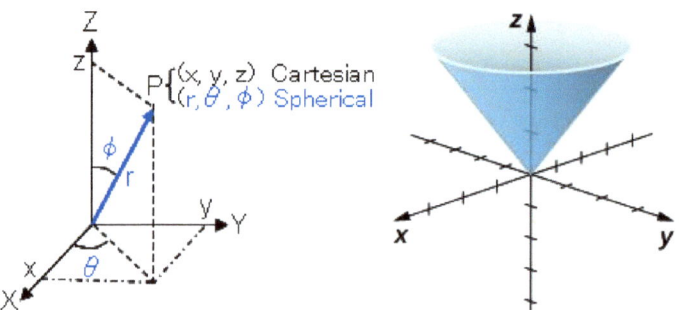

Figure 23 Cartesian and spherical coordinates (left). Cone (V_{cone} = 1/3 A_{base} h, circular cone V_{cc} = 1/3 π r^2 h), geometric 3D shape with an enclosed base which perimeter raise to a common apex (= vertex) in the Cartesian origin and height equal to the polar z-axis, erupted from the revolution of the radius of the base circumpherence, in which polygons can be inscribed in the shape of pyramids (right) - Image source: https://keisan.casio.com/exec/system/1359533867

Helixes are 3D cylindrical spiral coming from circles with x = cos(t) and y = sin(t) pulled along the z-axis, while, conical instead can come from the rotation of the Archimedean or Bernoulli spirals. Archimedes would throw the primordial corpuscle in the fluid column, marking the shah parabolically for raising according its intrinsic number of particles rejected by the elastic, tense, envelope on which it has been casted. The equiangular logarithmic **Bernoulli's spiral** polar equation is equal to:

$$r_{(t)} = e^t = e^{a\theta}$$

For example, the golden spiral has a = $[(1 + \sqrt{5})/2]^{2/\pi}$

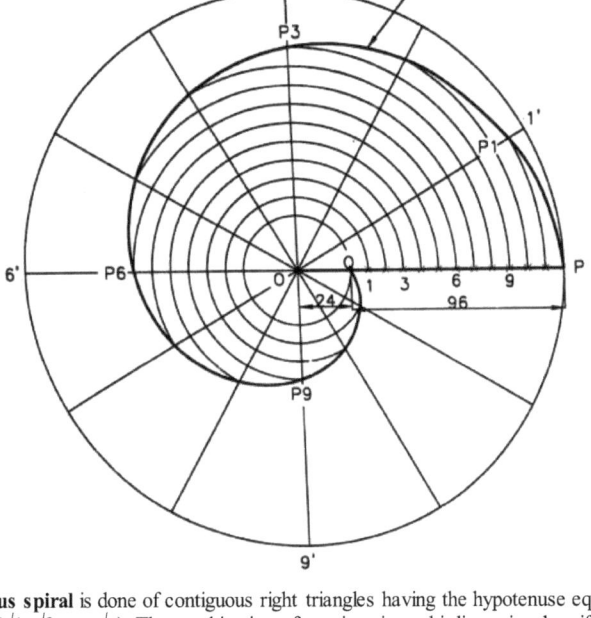

Figure 24 **Nautilus spiral** is done of contiguous right triangles having the hypotenuse equal to successive natural irrational ($\sqrt{1}$, $\sqrt{2}$, ..., $\sqrt{n}$). The combination of two imaginary bi-dimensional uniform movement at same time generated from the same particle, one linear the other circular, the radius and the circumpherence, respectively, permits describe the **Archimedean spiral** immortalizing the punctual corpuscle in motion, the cyclotrons, on the plane at constant angular velocity (ω) as plane waves, highlighting its position in eight points more the origin (O) at interval of angular time $t = 45° = \pi/4 = 7$ min 30 s, where touching the circle define the 4/turn parallels to the Cartesian axis, x∥ and y∥.

Leonardo Fibonacci (filius Bonacci) (c. 1175 - c. 1250) a Pisan mathematician, author of the Liber Abaci ('Book of Abacus or Book of Calculation') which popularized Hindu–Arabic numerals in Europe, famous for having defined the series $F_n = F_{n-1} + F_{n-2} = 1, 1, 2, 3, 5, 8, 13, 21, 34, 55, 89, 144, 233, 377,...$, n, where: $F_1 = 1$ $F_2 = 1$, with a *golden spiral ratio* $\cong 1.62$. The **Fibonacci series** could display the perfection ratio of the meeting, certainly shows biological symmetries such as the phyllotaxis, the petals arrangement, the number of possible ancestors on the X chromosome inheritance line at generation n (Hutchinson L., 2004) and the tetra-loops of *Orion*. Furthermore, his networking cubes with the vertices labelled with n-bit strings (Hsu, 1993) are used nowadays for interconnection between parallel or in series elements like the components of electronic circuit, the same that and E-Homo would know to be approximation of his, her anatomic systems, organ and tissues (e.g., pineal gland = diode). The petals of the sacral geometry **Flower of life** like the *Asteraceae* common daisy (*Leucanthemum vulgare*), perhaps, gave the name of the elastic coil of the Renaissance aforementioned, manifesting their beautiful perfection in unfolding themselves during the reproductive season, with the only difference that to ascent are not directly themselves, but their volatile fragrance; nevertheless, for each spring the magnitude of the displacing perturbation, increase with the distance from the sourcing receptacle. Actually, each curve inflection, corresponding to half-loop, otherwise, half-petal perimeter, is tangential with the perpendicular radius of the imaginary circumference in which the corolla is inscribed, and according the 'inverse-square law', generalized by Isaac Newton for any changing strength under gravity (g = 9.8 m/s² on planet Earth at sea level), even interplanetary thanks to Henry Cavendish gravitational constant empirical elucidation, the propulsive wave magnitude due to its roto-traslatory motion is inversely proportional to the square of the radius (intensity ∝ 1/ radius²). **Newton's law of universal gravitation**:

$$F = F_1 = F_2 = \frac{G (m_1 \times m_2)}{r^2}$$

Where:
G = Henry Cavendish gravitational constant = $6.674 \cdot 10^{-11}$ m³/ kg · s²
$m_1 = m_2$ = masses [kg]
r = radius [m]

Pythagoras of Samos (c. 570 – 495 BC) was a Magna Graecia teacher of the transmigration of the souls, believing, precisely, in the reincarnation, the resonating planetary motion, the homonymous trigonometry theorem, numerology (monad, dyan,…, various specific meaning associated to the number according the culture and, perhaps, for inducing a better allego-methaphoric integration in his students), tetracyts, standard 3:2 frequency tuning ratio for any interval, five regular solids, proportionality, Earth's sphericity and climatic zones, identification of the morning and evening star as the planet Venus. Point, line, plane, volume, then mathematics become physics, when the root is not more the Pitagoric diagonal, the factor that most significantly change the system become the inclination of the Galilean plane dividing the cavity from the particle, making a border in which that is visible motion. Certainly, you would agree that the displacement of the *Atman* from an individual to the another is just suitable for students yearning to be educated, differently from the transmigration of the *Nirvana Sharir* for which the lymphatic system of the death resonate within the one of the living praying for his, her rebirthing into him, her.

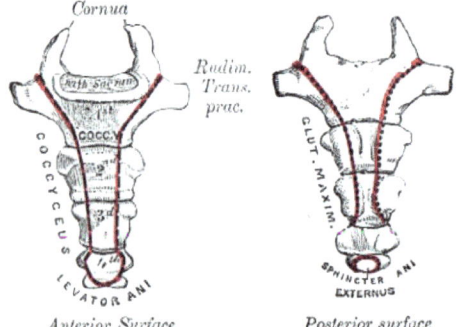

Figure 25 Coccyx, the set of the sleeping Kundalini where Sirus, Virgo and Gemini constellation, and the planet Mercury radiations inducing the raising of the cyclic electricity due to the cyclotrons orbiting in each chakra, through the neuromuscular junction along the vertebrae, tetra-loop after tetra loop in the Betelgeuse frequency resonance, till the Gate of the Soul, the pineal gland – Image source: Henry Vandyke Carter 'Anatomy of the Human Body' (1918).

The point coccyx *Sphincter ani* does not translate, just rotate on itself generating thermal energy, this triggers the kinetic energy to develop permitting the translational motion via the cones. After each turn, four new daughter circumpherences appear on the mother remaining connected to it in the origin (O) and overlapped in two points; the first positive (+) or A(x_A ; y_A) and the second, specular, negative (-) or B(-x_A ; -y_A), related among them by the following equation: B = - A. Generative counter-clockwise circular motion that could be associated to the cytological phenomena *mitosis* involved in the tissues growth and development or the microbial binary scission, considering the single cycle done of two following cellular divisions; at $t_0 = 0$ there is O cell. In the point P pass the parabola with vertex in the origin of the Cartesian axis and focus F (0 ; y_A) of equation $p(x_A) = ax_A^2$, where, a = parabolic constant = 0.5. Considering each contiguous point's sender and receivers in a continuous sequential pairing. The coccyx vertebras couple each other in succession, then its last the sacrum, this propagates the wave through the thorax that continue vibrating to the cervical. Couple the sacrum, this with thorax, finally, the cervical vertebras in differential steps along a sigmoidal spinal function, in which the logarithmic equiangular Bernoulli spiral latent phase is followed to the exponential then from an asymptotic stationary. The speedily Mercury escaped, would raise for elevating its path, oscillating, widening in the lumbar, and firmly translate describing a 3D cone with the vertex pointing the ground: the vinculum coccyx, the host of the sleeping Kundalini. The oscillating strings is in each point in superposition between elevating clock-wise (- *Pingala, Helios*) and grounding counter-clockwise (+ *Ida, Selene*), respect the imaginary axis passing between them, the *Sushumna nadhi*, but some of them, the singularity nodes peculiarly delimit the ellipsoids characterizing each chakra.

The unstable equilibrium, then chemistry inclination making in a dynamic equilibrium an isolated system is the sound, being the hearing the only human sense that cannot be rejected with the body, but with the mind, defensive and stimulus responsive. According to the **Newtonian principle of action-reaction forces** regulating, for which equality responses maintain a resilient state. Kirchhoff, Maxell, Ampere, Faraday, Tesla et al. regulating the attraction-repulsion duality of sub- atomic particles in an ElectroMagnetic field, expanded to the planetary cosmos thanks the isolation *in sensu stricto* that seen Henry Cavendish, totally alone and merged with the collective love, the manas on which each human is plugged into, define the gravitational constant aligning the infinitesimal small with the immense that see each other when time is annihilated in the instant in which time lose itself, vanishing instant after instant, leaving the space constantly acting, hence, leading the way with apparently null response. The coalescence of two black holes that gave origin two the Multi-Universe should have been happened logarithmically in the beginning under an unstable equilibrium slowly stabilizing, presumable at temperature proportional to T < 600 °C due to the Cosmic Wave Background (CWB) heating. The diffuse source of visible energy was cold and bluish, and enveloped all the other PAR radiation in concentric elliptical rainbow nuances when gravity wasn't shaping them as spherical, coherently with the convection heat transferring far-off the source, everywhere around the magnetic **Jupiter Singularity** (S), which, infinite gravitational field related triggered the full combustion of the enveloping medium: the hyper-cold primordial Multi-Universe. The mature Singularity had to be spherical, with layers of variable temperature divided by Ultra High Vacuum (UHV) gap, appearing bluish-violet due to primordial plasma lightenings depending of the selective composition of the ionized gases into it. Furthermore, it does not react with any other physical object, as nothing else existed apart the cold empty space. Physic laws and their constants like the speed of light in the vacuum are invariable under the perceptions of the relatives. Two events far-off distant can be related by a common, but mutual process as two quanta by an opposite rotation along their polar z-axis, otherwise, has two people of a couple that love themselves that even living at the antipodes keep exchange love among themselves via the Manas Sharir. The *Michelson-Morley luminiferous aether*, the Nirvanic void medium carrying light corpuscles support the electromagnetic wave deforming itself in all its point according the gravity filed overtime. Two right circular light cones with both vertex incongruent with the event origin (O) of the spacetime diagram, one extending into the future (t > 0) and the right in the past (t < 0). The radiation reaching the ground is stored in the past cone electrode counter placed right under the horizon O as a venerean *Daucus carota* (carrot) deep in the soil that can be observed just sweeping the upper layer of the Earth to emerge in the light of its future growth and development despite its great amount of energy is hidden in its own past. Common innate memories travel through light and are shared by every sentient being living on the planet Earth, perhaps, also in other astronomical roto-traslating synchronic objects. Hence, is strictly needed to catch all the probable radiations with any period from the quicker to the slower and *vice versa* for later emitting it in various connected devices. If the 2D circumpherences become 3D spheroids, the corpuscles, become disperse moving within them in independent rotation. The *electrons e⁻* would orbiting counter-clockwise along the arch. The spheroid Singularity rotate with synchronisms between their z-axis, which corner respect x-axis gives the time. The anisotropy that makes the e⁻ in a probabilistic volume is eliminated by the synchronism of the five multi-universe spheroids thanks to which is possible to localize the electrons in a dynamic Multi-universe, knowing contemporary both the position and the momentum overcoming the Heisenberg uncertainty principle. The electron orbiting would define a not ergodic pathway, on which symmetric diagonals with the common Cartesian coordinate would give four tangential points corresponding to its dial localizations. The proton (p^+) would be in the core of a **Hubble sphere**, along which cosmological horizon the electron would fall through the conical temporal eyes that connect the cardinal spheres with the proton Hubble sphere. Denser envelopes would contain less massive particles per volume occupied, four rotating bubbles around an original central, before in time at bottom of a wider containing light cone.

Figure 26 Xipe Totec the Aztec God related to the generation of the planet Jupiter with a rainbow conical hat that would depict the White Hole to the other side of the mirror emitting all the Sun spectra radiation since before the Big Bang.

Remember, there are two way in the cone, one transcendental ascensional for travelling forward in a superconductive spiral, and another slower drunk eradicating for celebrating the landing. **Conical spiral one, Marangoni chain the other**. An elliptical path with four nodes is described along the circular surface of this Dirac time cone, and its four loops corresponding to the intact bubble, due to the Dionysian *Marangoni effect*, immersed in it. The thermos-capillary convection phenomena permit to the massive particles transferring from the alkali metal liquid surface of the bubbles in its noble gases non-metal gaseous core.

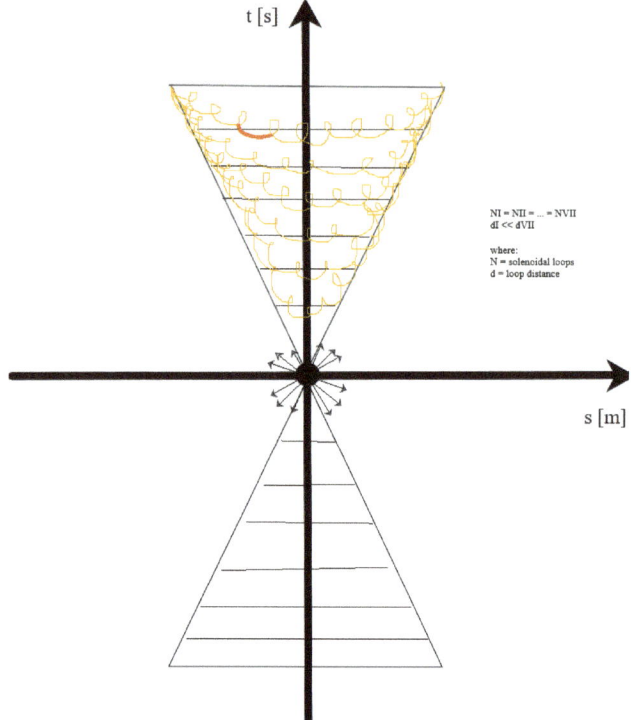

Figure 27 *Marangoni effect* of the waving corpuscle that gave origin to the Bubble Universe roto-translating against gravity along the event horizon, for later falling becoming the *Singularity* – Image source: © Copyright Antonio Silvestro, 2020.

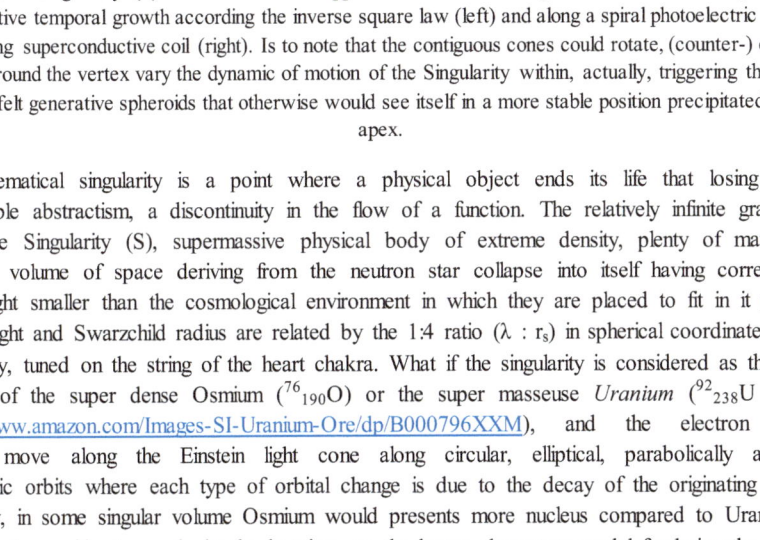

Figure 28 Singularity (S) at base of the overlapped Dirac bi-cone during its first moan rebirthing and its projective temporal growth according the inverse square law (left) and along a spiral photoelectric effect undergoing superconductive coil (right). Is to note that the contiguous cones could rotate, (counter-) clockwise (-, +), around the vertex vary the dynamic of motion of the Singularity within, actually, triggering the raising of the felt generative spheroids that otherwise would see itself in a more stable position precipitated to the apex.

A mathematical singularity is a point where a physical object ends its life that losing itself in undefinable abstractism, a discontinuity in the flow of a function. The relatively infinite gravitational spacetime Singularity (S), supermassive physical body of extreme density, plenty of matter in a constrict volume of space deriving from the neutron star collapse into itself having corresponding wavelenght smaller than the cosmological environment in which they are placed to fit in it properly. Wavelenght and Swarzchild radius are related by the 1:4 ratio ($\lambda : r_s$) in spherical coordinates for any singularity, tuned on the string of the heart chakra. What if the singularity is considered as the atomic nucleus of the super dense Osmium ($^{76}_{190}O$) or the super masseuse *Uranium* ($^{92}_{238}U$ \$ 78/kg https://www.amazon.com/Images-SI-Uranium-Ore/dp/B000796XXM), and the electron orbiting around move along the Einstein light cone along circular, elliptical, parabolically and later hyperbolic orbits where each type of orbital change is due to the decay of the originating nucleus? Certainly, in some singular volume Osmium would presents more nucleus compared to Uranium, but this last alone without any doubt the heaviest, so the better elementary model for being the chemical singularity that gave origin the Multi-universe that we all know with fringe is done by the Big Bang. This actinide element has 6 available electrons ($1s^2 2s^2 2p^6 3s^2 3p^6 3d^{10} 4s^2 4p^6 4d^{10} 5s^2 5p^6 4f^{14} 5d^{10} 6s^2 6p^6 5f^3 6d^1 7s^2$) for orbiting around the light cone, and its radioactivity unstability, decay emitting alpha rays is congruent with the thoughts fall from a conic section orbit to another. Furthermore, its half-life maximum pick of $4.5 \cdot 10^9$ years makes it a perfect candidate to be the 'element singularity' on the border between past and future of the Multi-

Universe at the turn of the rapid matter expansion of the extremely high density and temperature singularity which according to current cosmological theories marked its genesis. Black from the upper, White Hole from the lower the nuclear singularity supported self-propagating nuclear chain reactions. Currently, hydrogen and helium formed when the Whole cooled down, but it was already in the future cone that happened just $380 \cdot 10^3$ years ago. The discontinuity laying in the hyperspace of the present corresponding to the point of time t = 0 has to be characterized by the presence of the only Uranium, extremely hot, massive and bending the elastic gravitational field, precisely, oscillating on its network in the past through the presence of the future continuously, till stopping in a determined special point: the event O done of t ; s = 0 ; y.

The supermassive element Uranium appear as a silvery white malleable, ductile, electropositive metal proper for being the plus event O of very high density (ρ = 19 g/cm^3), high energy $_{235}$U (E = 20 TJ/kg), characterized by fluorescent properties for this element would have been triggering the massless bosons photon release illuminating the dark matter of the white hole. Natural Big Bang and artificial nuclear bomb are destructive explosions, both leading a universal peaceful creation. The glowing uranium singularity under the UV quanta released by itself helically raising, roto-traslating along the walls of the light cone, inducing the fission that led the releasing of beta particles, electrons or positrons filling the future of the Multi-Universe. A fair amount of these sub-atomic particles balancing the ascensional orbiting in the light cone of the future in which we are living currently. Rapid neutron capture process is the phenomena behind the production of the two extant primordial uranium isotopes, ^{235}U and ^{238}U metabolizable, for example, *Shewanella putrefaciens*, the proteobacterium *Geobacter metallireducens*. The most important oxidation states of uranium are greenish and yellowish uranium (IV) and their two corresponding oxides, respectively, uranium dioxide (UO$_2$) and uranium trioxide (UO$_3$).

The Universe change within space (x, y, z) and time (t), while the Multi-Universe, according the **String Theory**, varies within 11 (M) or 12 (F) dimensions. In agreement with the F-Theory, a Multi-Universe could be done by the combination on three Universes on the Cartesian axis, or four different Spherical spatial coordinates with a common time line for which would be possible to overcome the Heisenberg uncertainty principle calculating simultaneously, position in one of the four spheres radiant corresponding, particle mass, radial velocity, while time on the common original sphere. For example, a corpuscle of mass m settled in O(0,0), would move along an Archimedean spiral pathway with an *overall velocity* done of the sum of the five sub-speeds of the particle in the circumpherences (o, a, b, c, d).

$$v = \sum_{i=o}^{d} v_n = v_o + v_a + v_b + v_c + v_d$$

Chakra strings works in three dimensions, describing spring that could be theoretically limitless (∞), deterministically stretching upward till a stabilizing value, this last harmonically calculable using the opposite deforming, stressing force defined by *Robert Hooke* (F = - k · x) in the XVII century, accelerated mass training respect the reference gravitational field or in interaction with one or more compressing massive physical objects. Each energetic wheel would contain explosive power variable depending on environmental pressure and gravity. **Entropy of a low-density monochromatic radiation** like the string of *Anahata chakra*, varies with volumes in the same manner as the ideal gas or a dilute solution (Herr. Albert Einstein 'Ann. Phys. 17, 132', 1905).

$$S - S_0 = (E/\beta v) \ln\left(\frac{V}{V_0}\right)$$

As witnessed and described previously by P. Lenard for continuous sunlight propagated through the space and late quantum-mathematically corroborated by A. Einstein the **Photo-Electric phenomena**, would be the process for which the incident light triggers the electrons motion within a conductive physical object like the pineal gland, related to the following equation:

$$\Pi q_e + E_c \geq R\beta v \Rightarrow \Pi \geq (R\beta v - E_c)/q_e \Rightarrow \Pi \geq (R\beta v) - F_c [2\pi \sqrt{(a^2 + b^2)/2}]/q_e$$

Where:
Π = potential [V]
q_e = electron charge [q]
E_c = emissive work of the cyclotron = $F_c \cdot x = F_c [2\pi \sqrt{(a_2 + b_2)/2}]$, a>b
R = universal gas constant = 8.3145 J/mol
$\beta = 4.866 \cdot 10^{-11}$
v = frequency [Hz]

The Kundalini vortex ascending or descending in time, widening or reducing in volume along the *Orion* vertebral column enveloping Bi-Dirac cones chakra-representing, constellation and planets. Nuclide spherical harmonic can represent geometrically any **nuclide as a locus** (Latin: *'place, location'*), geometrically a set of points on a plane defined by determined conditions such as the conical sections parabola, circle, hyperbola, ellipse, precisely, orbiting along a circle clockwise (+) for later falling under gravity to the core releasing with the impact its content, protons and electrons. Mass left from the massive would let to the generation of the not having one, stripped-out electron raising along the gravity walls of the cones on helical counter-clockwise (-) pathway as excitons emitting photos, **quantum leaping colouring each chakra** differently. Seven integral area corresponding to the seven visible colours of the *Photo Active Radiation* (PAR), and the corpuscles of related nuances that in the positive direction up-outward for convention would decaying with each spiral turn, transmuting, losing enveloping matter, but acquiring energy as rockets ready to take-off. As names is the inner condition of any physical body, even the singularity is not firm trying to escape the gravity network spinning, accelerate moving along the perimeter of the light cone leaving decaying shah perpendicularly to its path colouring the circumpherences, in the first turn it lost matter, hence, decrease its radius, what permit is helical raising movement along the walls of the light cone, loop by loop colouring it from red to violet. The ending 7 main nuanced light cone. The component x and y of the light cone are geometrically infinite, but has light is characterized by an ascending energetic value, when is reached the circular halo accretion disk the red particle, the cone will be not more done of visible PAR light, but invisible radiation of lower energy till the ELF.

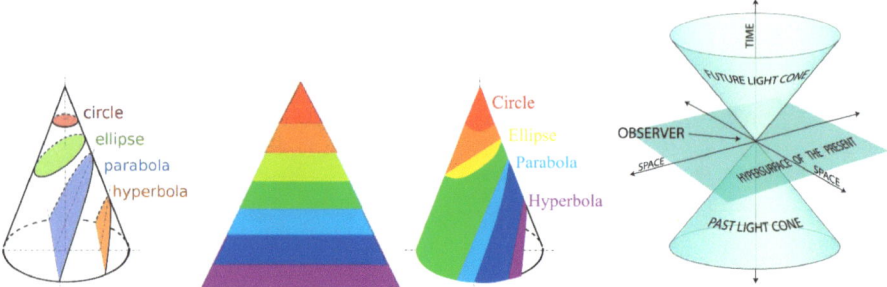

Figure 29 Conic sections (circle, ellipse, parabola, hyperbola) (top left). Colourful conical spirals, curves $z = z_0 + mr(\phi)$ on a right Einstein circular light cone $m^2 = (x^2 + y^2) = (z - z_0)^2$, $m > 0$, would describe the harmonic *ElectroMagnetic* (EM) radiations series from the fundamental singularity (F note) to the present hypersurface $x = r(\phi) \cos\phi$ and $y = r(\phi) \sin\phi$. *PhotoActive Radiation* (PAR) and triangle (top centre) and conic section (top right). N.B. barycentrum passing on the height between orange and yellow (bottom). Past and future light cones between the present hypersurface (bottom). – Image Source:
https://commons.wikimedia.org/wiki/User:LucasVB,
https://commons.wikimedia.org/wiki/User:Magister_Mathematicae,
https://en.wikipedia.org/wiki/User:K._Aainsqatsi - Image source: © Copyright Antonio Silvestro, 2020.

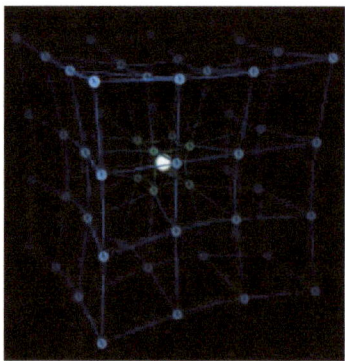

Figure 30 Curved continuum spacetime with the Singularity in the centre.

As light is continuously diffused in reality, but can be imagined as pointed for better describing the complexity of the Universe in which it flows, so space and time are merged in an unbroken dimensional whole, the Hermann **Minkowski spacetime** combination of three-dimensional Euclidean space and time into a four-dimensional pseudo-Riemannian manifold, where difference of potential among two bent locus is always present, which event interval is independent of the inertial frame of reference in which they are recorded, for precisely defining a physical object or the environment in which it is placed in the way most probably congruent with the one observed by the absolute. Infinite positions lost in instantaneous time ranges and shapeless spaces can be acquired by objects involved in continuous transitions, each defined by a set of coordinates x, y, z and t, aligned on the common scaling reference speed of light through the vacuum ($c = 3 \cdot 10^8$ m/s), placed in linear succession, one after another in a probabilistic pathway. All the spatial dimensions x, y and z depend on time t, each of the objects in the network could be the reference frame, being the emitted light from each of them of equal value, but is needed to define the flow direction for assessing any system in a reproducible way once it has been choose switching coordinate to another relative frame can be easily done with the magnetic *Lorentz transformations*. Measures comes after observations and are determined by previous knowledge integrated by the new acquired data properly processed, but there is no more delay when is light the witnessed at least in a relativistic vision where no hypothetical tachyons are taken in consideration. The spacetime would bend the Singularity proportionally to the strain **tensor** of the gravity field, symmetrical between the two holes, electron acceptors like the *Mulhadhara* and the *Shahashrara* chakra, is the **product** $\otimes$ of the position ($|\psi(t)\rangle$) and spin state vector ($|\xi(t)\rangle$) at the intersection of the Kirchhoff network, for example, between the White and Black Holes equal to:

$$|\psi(t)\rangle \otimes |\xi(t)\rangle = \sum_{s_z} \int d^3\,r\,\psi(r,t)\,\xi(s_z,t)|\,r\rangle \otimes |s_z\rangle$$

Now the query is, it has been before the Singularity or the cone? Well, you may now that as between the egg or hen of *Gallus gallus*, there have been the reptilian ovulum, so the enveloping Universe Bubble.

As according the thermodynamics law energy doesn't create nor destroy, anyone may argue that recreating the whole Bubble Universe in vitro would make you able to conserve idealistically all the existing energy. Currently, energy stored in the chemical bonds of biological ion-exchange redox membranes and macromolecules such as glucose polysaccharides carbohydrates such as the animal glycogen and the vegetable starch, (bio)fuels thermally expanding, fluids flowing through empty channels, electromagnetic field of (super)capacitors or cryogenic superconductive coils.

As Nikodem Poplawski of Indiana University in Bloomington would agree, a Wormhole may have preceded the Big Bang giving it the conditions for being the cosmological event onto which the whole Bubble Universe lays on. **Artificial wormhole** has been made using a sphere as approximation of the primordial Universe confines done of an outer layer of shielding ferromagnetic nickel (Ni) mu-metal, an inner of superconductive Yttrium Barium Copper Oxides (YBaCuO$_2$), inducing magnetic field according the Lenz-Faraday law of induction, filled by freezing liquid *Nitrogen* at $T_{N2(l)}$ = - 195.79 °C encapsulate into the bilayer, used for freezing, drying, and as cryogenic in biomedical devices and astronomy CCD camera, with a centre diametral cylinder done of an enrolled mu-metal lamina (https://www.scientificamerican.com/article/magnetic-wormhole-created-in-lab/) bent by the Minkowski spacetime.

Figure 31 Artificial wormhole – Image source: Jordi Prat-Camps / Autonomous University of Barcelon, 2015.

Standard Reduction Potentials in Aqueous Solutions at 25 °C

Oxidizing Agent			Reducing Agent	Reduction Potential (V)
F_2	+	$2e^-$ →	$2F^-$	2.87
H_2O_2	+	$2H^+ + 2e^-$ →	$2H_2O$	1.78
MnO_4^-	+	$8H^+ + 5e^-$ →	$Mn^{2+} + 4H_2O$	1.51
Au^{3+}	+	$3e^-$ →	Au	1.50
Cl_2	+	$2e^-$ →	$2Cl^-$	1.36
O_2	+	$4H^+ + 4e^-$ →	$2H_2O$	1.23
$Cr_2O_7^{2-}$	+	$14H^+ + 6e^-$ →	$2Cr^{3+} + 7H_2O$	1.23
Br_2	+	$2e^-$ →	$2Br^-$	1.07
NO_3^-	+	$4H^+ + 3e^-$ →	$NO + 2H_2O$	0.96
Ag^+	+	e^- →	Ag	0.80
I_2	+	$2e^-$ →	$2I^-$	0.54
Cu^+	+	e^- →	Cu	0.52
O_2	+	$2H_2O + 4e^-$ →	$4OH^-$	0.40
Cu^{2+}	+	$2e^-$ →	Cu	0.34
$2H_3O^+$	+	$2e^-$ →	$H_2 + 2H_2O$	0.00
Pb^{2+}	+	$2e^-$ →	Pb	−0.13
Sn^{2+}	+	$2e^-$ →	Sn	−0.14
Ni^{2+}	+	$2e^-$ →	Ni	−0.26
Fe^{2+}	+	$2e^-$ →	Fe	−0.45
Cr^{3+}	+	$3e^-$ →	Cr	−0.74
Zn^{2+}	+	$2e^-$ →	Zn	−0.76
$2H_2O$	+	$2e^-$ →	$H_2 + 2OH^-$	−0.83
Mn^{2+}	+	$2e^-$ →	Mn	−1.19
Al^{3+}	+	$3e^-$ →	Al	−1.66
Mg^{2+}	+	$2e^-$ →	Mg	−2.37
Na^+	+	e^- →	Na	−2.71
Ca^{2+}	+	$2e^-$ →	Ca	−2.87
Ba^{2+}	+	$2e^-$ →	Ba	−2.91
K^+	+	e^- →	K	−2.93
Li^+	+	e^- →	Li	−3.04

Increasing Strength of Oxidizing Agent ↑   *Increasing Strength of Reducing Agent* ↓

FLINN SCIENTIFIC, INC
"Your Safer Source for Science Supplies"
© 2007 Flinn Scientific, Inc. All Rights Reserved.
AP7041

Standard Reduction Potential ($E^0 = \Delta V^0$) in aqueous solution at T = 25 °C
(https://www.av8n.com/physics/redpot.htm,
https://www.phoenixcontact.com/online/portal/pi?1dmy&urile=wcm%3apath%3a/pien/web/main/products/subcategory_pages/Power_supply_units_P-22-03/fe33ddbe-218c-4e46-97b4-6f12d82de835)
according to IUPAC definition, is the electromotive force of a electrolytic cell, that can be used also in the calculi concealing the teleporter. Half reduction potential:

$$2\,H^+_{(aq)} + 2\,e^- \rightarrow H_{2(g)}$$

SHE 4.44 ± 0.02 V at T = 25 °C, but declared $\Delta V = 0$ V

$$O_{2(g)} + 4\,H^+ + 4\,e^- \rightarrow 2\,H_2O$$

$$E°(\Delta V) = +1.23 \text{ V}$$
$$\text{Electric field} = E_{cathode} - E_{anode} - \eta$$

Where:
η = ohmic resistance losses, activation switch and mass loss term

The charged elementary particles, quarks and leptons, in the UHV would be glued differently by the Black/White Holes **electrodes** (ancient Greek: ἤλεκτρον élektron amber and ὁδός hodós way term coined Michael Faraday) depending on surface area and Nafion porosity (ø = 40 Å), walls capacitance (C), electrical conductivity of the reference catalyst (σ_{Cu}), stability, durability, accessibility and costs, certainly, composing them with unbelievable power supply generated and stored continuously in a loop feedback based on the origin of the whole Universe.

Inorganic catalysts (https://www.ptable.com/) are typically various small ligands coordinated in a dative bond, both the electron of the covalent bond is donated from just one of the two atoms in interaction, with a transition ionized metal centers (e.g., Ag, Ni, Au, Pt, Mo, Cu, Zn, Hf). Platinum ($_{28}$Ni = 1 €/g, $_{78}$Pt = 150 €/g, 15 €/cm, $_{79}$Au = 350 €/g https://www.novaelements.com/platinum/#cc-m-_product-13484589627) is among the best for catalyzing the decomposition of hydrogen peroxide (H_2O_2) into water (H_2O) and oxygen (O_2) and is also used in fuel cells for the *Oxygen Reduction Reaction* (ORR). To note that transition d-group metals, cheaper than alkaline metals (e.g. $_{37}$Rb = 200 €/g ΔV_{Rb}= - 1.97 V), can be used for coating the Cu core for incrementing their performance (e.g. Ni/Mo = 0.4/cm^2 with a non-significative loss in Hydrogen PRoduction (HPR 2 instead of 2.3 €/day/m^3). **Carbonaceous organ catalysts** will lack metal centers and have many carbon-carbon and carbon-hydrogen bonds (e.g., carbon cloth, graphite felt (3 x 3 cm)). P_{Ti} = 24 W/m^2 and $P_{grafite}$ = 26 mW/m^2. Ti/Ru alloy increase HPR and methane CH_4PR 4 and 13-fold, at 1.4 < ΔV < 1.8 V, respectively.

Table 1 Wormhole *in vitro*

Materials https://www.ptable.com/	Expensivest	Cheapest
Metallic conductive strip busbars	^{47}Ag, YBaCuO$_2$	^{29}Cu
Finger's radiation receivers	^{47}Ag	^{13}Al
Laminated insulator wafer	^{14}Si-, ^{22}Ti-, ^{77}Ir-, ^{238}UO$_2$ aerographene	Organic, c-Si, ^{28}Ni, ^{50}Sn, ^{49}In^{50}Sn^{16}O, EVA sheets
Transparent Conducting Oxide (TCO) metal Anode – White Hole donating (P +)	N-type semiconductor Arsenic (^{33}As), Phosphorous (^{15}P), Nitrogen (^{7}N), Antimonies (^{51}Sb), Bismuth (^{83}Bi), Moscovium (^{115}Mc) (valence = + 5, 1 free excess e⁻), Zinc (^{30}Zn), hydrogen (H$_2$)	SS (^{24}Cr, ^{6}C, ^{42}Mo), ^{29}Cu
Dye	^{44}Ru, Ti$_2$O, U$_2$O, SiO$_2$	^{12}Mg (isolated from chlorophyll)
Cathode – Black Hole electron accepting (N -) (ø = 2 nm) in regenerative electrolyte solution of I⁻/I^{-3}, KB, or LiB	P-type semiconductor ^{14}Si, ^{78}Pt, ^{32}Ge pure holes (valence = + 4), polypyrrole-coated *Carbon Nano Tube* (CNT)	CoS, ^{5}B, ^{13}Al, ^{31}Ga, impurity holes (valence = + 3)
Substrate	hydrogel	Pectinaceous fruit peels grinded into jam
Wires	^{47}Ag, TiO$_2$, UO$_2$, SiO$_2$	^{29}Cu
Lenses	B$_2$O$_3$	B$_2$O$_3$

Mirrors	^{47}Ag	^{13}Al

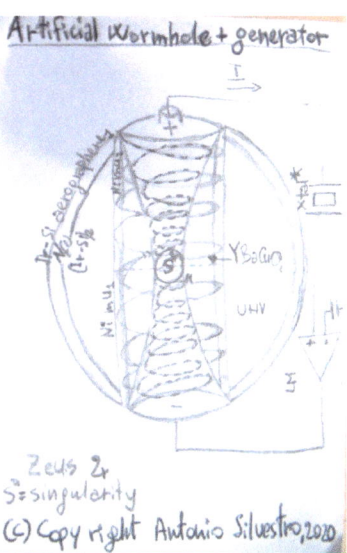

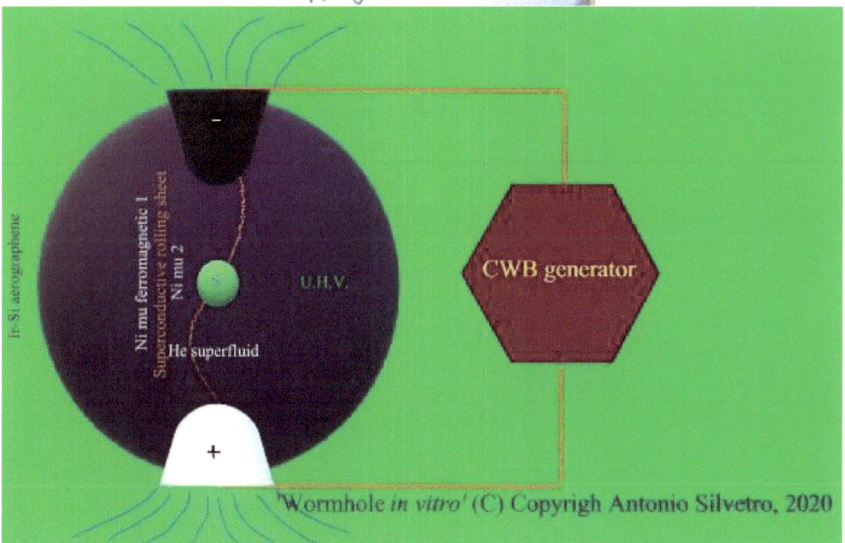

Figure 32 *'Artificial Wormhole (Jupiter Singularity) + CWB generator'* pencil sketch (top) and Paint 3D (bottom): Ir-Si aerogel insulating Ultra High Vacuum (UHV) sphere into which a superconductive rolling sheet within a bilayer sandwich of ferromagnetic Mu-metal (Ni/Fe), envelopes the cylindrical super capacitative double cell into which the Singularity, containing the among most close to LUCA descendant, *Cyanobacteria,* in submerged into fused hydrogens (H$_2$), alpha particles (4-He) helium superfluid medium through which a superconductive helical flow happen from oxidising and fusing cathode Black Hole (-) to reducing and fissing anode White Hole (+) – Image sources: © Copyright Antonio Silvestro, 2020.

Spinning around and rotating upside down a tin can containing a singular plastic bead, would let you feel what God did in the beginning with our Bubble Universe under vacuum, the owner of that inestimable colorful corpuscle that may bring you back on time to where you found it, before the beginning itself, before creation and destruction, triggered the same wave oscillating within your surprise heart finding it. Fill it with grinded XPS wet of fuel, now, make it, it will resemble Napalm of the Vietnam war, a lot of energy, many particles and among the many your own bright ball. Would you be able to find it back once putted into? Would it merge within the plastic medium? So mesmerizing silicon insulated sphere from which 360° a lonely microbe looking out seeing the destruction of the ignition of the combusting fuel, hot and dense, extremely from one side already. So, hurry up, hold it from the other side where your hand is not insulating it generating plasma lightening inward. Zeus in your thunder generating fist, with clepsydra of the father Cronus into. Sparks up and down avoided to come out by the electron aero-graphene plates. The least dense solid (ρ = 160 g/m^3), even lesser than helium (He), would act has the solar flares coming out from the Sun surface, in knots, untangled along the cable connected to CWB generator. The was before the wave of the cylindric tank? The sperm of the father or the ovum of the mother? Certainly, before the last, inferior, less developed, merely, more stupid, always in look for understanding what's that inside. A ball, girl, just a ball as your enclosed in the scrotum, but you are still here trying to let goes out that cute microbes, freeing them from the core fire, that scary you too, man, so huge and arrogant in looking for clarify what do with the whole Bubble Universe. Certainly, you may understand that the vibration generated by the CWB would favor the motion of the bio-bead, sealed, protected by the Si gaseous gel window. Explosions, contained by Ni-Mu ferromagnetic shielding EM field, would scary that minuscule particles, alone in the Singularity, one only has her. Do you not it may be size one integer, only one, enclosing itself in the origin where the four wings of the Nafion PEM butterfly would generate for the Coriolis effect creation to the other side where destruction is happening, pulse after pulse, beat after beat within your heart. The crystal oscillator, that piezoelectrically would generate the wave displacing the loved LUCA descendant along the helical current tangentially along the cortical walls of the clepsydra that would bring you back on time, on the time of creation of the whole, in your hands with a coke. Please, drink it. Do let it sip out losing its energy, the energy of that subatomic corpuscle that just a Tokamak may contain, then nuclei with traits of lepton, that unknown at the origin that make see us see the mirror as a beating pair of *Lepidoptera* wings. Would we be able to save that *Cyanobacteria* from being oscillating endless, or even being burned by them extremely hot flames that you may have seen looking into a Vulcan, resonating as the clepsydra of Cronus going out sip after sip drinking the $CaCO_3$-based black solution, living us see just the hole of this color. The alkaline shell dissolved would you let understand that the same walls of the Singularity may merge with the lightest acid medium within it, neutralizing each other, in an energy generating exchange of protons flowing through the Nafion.

'Wormhole *in vitro*' © Copyright Antonio Silvestro, 2020

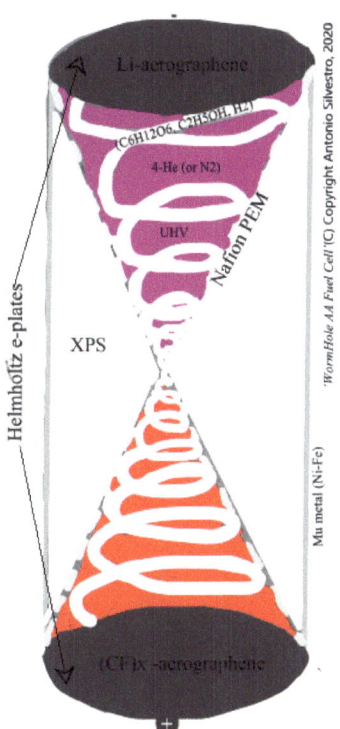

Figure 33 *'Wormhole AA Fuel Cell'* in which biofuels flow such as glucose ($C_6H_{12}O_6$), bioethanol (C_2H_5OH) or hydrogen (H_2 hydrogen automotive storage m = 50 kg, V = 50 L, ρ = 1 kg/L, t = 2.5 min, 300 €) into an empty thermo-insulated dielectric glass [sand (silicon dioxide, SiO_2) + limestone (calcium carbonate, $CaCO_3$) + sodium carbonate (Na_2CO_3)] helix filled with cooling down 4-helium superfluid (4-He) or cheaper fluid nitrogen (N_2), surrounded by Ultra High Vacuum (UHV), inscribed in an Einstein-Dirac double cone of Nafion cation exchange membrane (PEM) walls, enclosed into permeable ferromagnetic shielding Mu metal (80 % Ni, 20 % Fe) cylinder filled with thickening Extruded PolyStyrene (XPS), from the reducing White Hole anode done of the most positive gaseous ion, fluoride, filling the hexagonal meshes of the graphene aerogel [Li-$(CF)_x$ BR-code battery C ≤ 5 kF, $\Delta V_{F(g)}$ = + 2.9 V, 425 Wh/kg, 1 kWh/L] to the oxidizing Black Hole cathode (most negative ΔV_{Li} = - 3 V), would store energy between the two opposite super-capacitative Helmholtz electrostatic plates in an efficient and reliable way for charging plugged devices like the Cosmic Wave Background Genset (CWBG), that itself may work as motor spinning the clepsydra recharging the source AA type Hydro-Electro-Magneto-Bio-Battery (HEMBB optimal values: - 20 < T < + 70 °C, 1 < ΔV < 4 V, 0.5 k < recharges cycles 1 M, 0.1 < C < 470 F, 1.5 < E/m < 265 Wh/kg, 0.3 < P/m < 1.5 W/g, weeks < $t_{self-discharge}$ < months, efficiency 95 %, 3 < $t_{lifetime}$ <10 years) – Image source: © Copyright Antonio Silvestro, 2020.

Helios irradiating

Figure 33 Sun and planets size comparison – Image source: CC BY SA 3.0 LSM Pascal
http://www.lesud.com/lesud-astronomy_pageid81.html

The Sun (ø = 1.39 Gm, 73 % H_2 + 25 % He + other elements O_2, Ne, C, Fe 2 %, d_{Earth} = 1 AU = 1.5 10^8 km, B_{Sun} = 0.15 mT ≈ 10^4 · B_{human}) is a relatively light and bright star, a sphere of hot plasma generated since the Big Bang nucleosynthesis during the first t = 20 min of the whole Bubble Universe, irradiating the energy of the nuclear reaction proton-proton p-p chain (H_2 -> He) happening in its core (20 %, ρ = 150 g/mL, T = 16 MK, E/V = 265 kW/L), that in about 5 billion years would fuses all its hydrogens in the core becoming a red giant and cooling down degenerating into an even denser white dwarf losing completely its shell in the last 75 kilo years. The Heliosphere Magnetic Field (HMF) has an Archimedean Parker spiral shape in the interplanetary space filled by the Sun of astonishing beauty. The Sun, the first renewable environmental source of the Solar System (SS), at temperature of T = 15 · 10^6 °C composed of 91 % hydrogen (H) atoms compressing themselves to fuse their nucleus covalent forming the noble gas helium (8.7 %), 3.6 × 10^{38} protons (hydrogen nuclei), or roughly 299 · 10^6 metric tons of hydrogen, are converted into helium nuclei every second releasing energy at a rate of 3.86 x 10^{26} J s. While heating up or relatively moving under an electromagnetic field can induce, reciprocally, orthogonal electrical storm according to the Michael Faraday and Charles Carlton Maxwell, empirical laws and mathematic equation, respectively, and break apart the H-atoms into charged ions, turning into the atmosphere, the self-conductor of the magnetic Earth, into plasma giving it a shape, travelling to the surface as random bouncing around packets of energy-photons.

The **Solar Power Spectrum** $[S_{xx}(v)]$ is the function that describe the signal emitted by the discrete package of energy – quanta, variable along the time line x(t), dependent on their characterizing frequencies. The core thermonuclear ElectroMagnetic waves emitted by the Sun are absorbed by itself as the matter strep-out form the crust of Mars roto-traslating around it, but for engineering purposes the solar radiation embrace all the ElectroMagnetic Spectrum from Extremely Low Radio Frequency ($v_{ELF} = 3$ Hz : $E_{ELF} = 12.4 \cdot 10^{-15}$ eV = 12.4 feV) to Ionizing [$v_{\gamma\text{-Rays}} = 300 \cdot 10^{18}$ Hz = 300 ExaHz (EHz) : $E_{\gamma\text{-Rays}} = 1.24$ MeV]. Total Sunlight radiation at Earth ground level: InfraRed (IR = 55 %) + Photo Active Radiation (PAR = 42 %) + Ultra-Violet (UV = 3 %). The radiation of your brain and heart do not make distance to this statement, as we all are oscillating matter since the sleeping powerful coccyx.

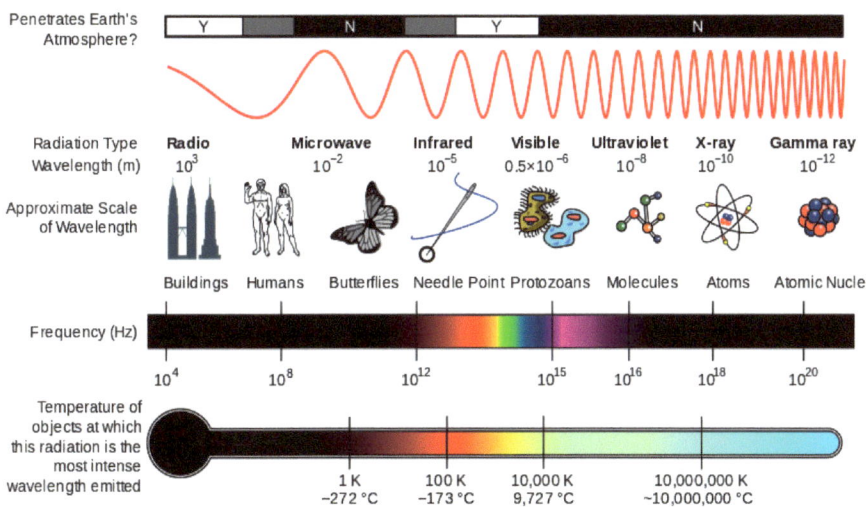

Figure 34 ElectroMagnetic spectrum - Image source: http://nasa.org/

Extra-terrestrial energy of the radiation:

$$E_{ext} = E_{sc} \left[1 + 0.03 \cos 2\pi \frac{(dn - 3)}{365}\right]$$

Where:

E_{sc} = *Solar Illuminance Constant* represents the invariable incident light onto a point, segment, surface or volume, in this last case is actually called *flux density* = $128 \cdot 10^3$ lx [lux = radiant flux/ m² = lumen/ m²]

During an ordinary day it would be equal to:

$$E_{dn} = E_{ext} e^{-æm} = E_{ext} e^{-æ \int q_{air}}$$

'Wormhole *in vitro*' © Copyright Antonio Silvestro, 2020

Where:
æ = atmospheric extinction, elastic flash of astronomical objects done of absorbance and following emissive scattering bring down the *Real Solar Illuminance Constant*: $E_{Rsc} = 10^5$ lx
m = mass of the air in the narrow tunnel in which the ElectroMagnetic radiation flows
q_{air} = 1.225 kg/m³, at sea level and T = 15 °C.

Earth-Sun *Astronomical Unit* (AU): 1 = E_{ext} : 1367 W/m² : P_{ground} : 1 kW/m²

Luminous flux ≤ radiant flux (emitted, reflected, transmitted or received).

Radiation a dual aspect wave-corpuscle transmitted through the full medium or empty space classified into ElectroMagnetic, Thermos-Nuclear, acoustic or gravitational tent to occupy all the volume in which they flow with or without elements remaining relatively more stable in the vacuum on the border of the *Solar System* (SS). Their intensity is inversely proportional to the square distance of the emitting-absorbing punctual source (E/r²), where: r² correspond to the ellipsoid half cavity area - 'hemi-ellipsoid', hence, the **'Holistic Radiation Equation'** (HRE) [J/m²] is equal to:

$$H = \frac{E}{r^2} = \frac{E}{\pi c^2} + \frac{\pi ab}{\sin(\varphi)} \left(E(\varphi, k)\sin^2(\varphi) + F(\varphi, k)\cos^2(\varphi) \right)$$

Where:
$\cos(\varphi) = c/a$
$k^2 = [a^2 (b^2 - c^2)]/b^2(a^2 - c^2)$, $a \geq b \geq c$

F (φ, k) and E (φ, k) = **elliptic integrals** of first and second kind, respectively $= \int_c^x Q(t, \sqrt{P(t)})\, dt$

$$F(\varphi, k) = \int_0^{\frac{\pi}{2}} \frac{d\theta}{\sqrt{1 - k^2 \sin^2 \theta}} = \int_0^1 \frac{dt}{\sqrt{(1 - t^2)(1 - k^2 t^2)}}$$

$$E(\varphi, k) = \int_0^{\frac{\pi}{2}} \sqrt{1 - k^2 \sin^2 \theta}\, d\theta = \int_0^1 \frac{\sqrt{1 - k^2 t^2}}{\sqrt{1 - t^2}}\, dt$$

Φ = Amplitude = 90° = π/2 and x = 1
K = ellipse eccentricity = sin(β/2) = 1
β = opposite angle of the foci distance, inscribed in the vertex of the equilateral rectangle
Q = rational number = a/b
P = polynomial of 3rd or 4th degree, e.g. $(ax + by)^4$
a, b = Z = integers
c = constant

From the aforementioned would be easy to understand that all the physical quantity related to corpuscles and waves can be calculated, exempli gratia, the *Holistic Power* = H/t [J/m² s = W].

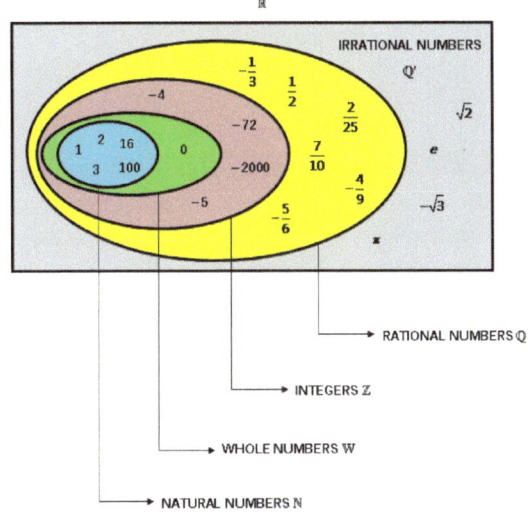

Figure 35 Numbers.

The main difference between photoelectric and photovoltaic effect is that in the first the electrons are emitted into the empty space, whereas, in the second the negatively charged sub-atomic particles flows into matter of relatively high density that the void. PhotoVoltaic Effect (PVE), the creation of voltage and current in a material upon light exposure, deducted by A. E. Becquerel in 1839, precisely, as a phenomena occurring when two inorganic catalyst plates (e.g. Cu, Pt, Ni, Si or Au) in a salty ionizing hydro-solution are exposed to ElectroMagnetic radiation emitted by the Sun or an artificial neutron sources (UV 250 < λ < IR 2500 nm, $\lambda_{sea\ level}$ = 1500 nm), and perhaps, also assessable on the accretion discs of the Black and White Holes of a wormhole surrounded by CWB, but in its case in an autocatalytic retroactive positive feedback in which the frequency absorbed by the black trigger the emission of the white cone.

Sandy oceanic and sea shores, containing SiO_2 nano-beads also derived from diatoms frustules, brought on surface from the backdrop, and variably concentrated salt like in the Mediterranean sea [NaCl] = 35 mg/mL, undergoes for a limited time range the PVE, precisely, when ones/twice a day a thin layer of moon tide cover them, hence, the photon-flux (Φ_q) from the Sun is converted into superficial vertical tension {ΔV [V]} and horizontal electric current {I [A]}, having its maximum irradiance, solar radiation flux density {SI [W/m²]} value in the equator at noon, where, the light rays perpendicularly falling onto water mirrors are fully converted in electricity as the diffracted solar radiation is at α = 90° with the incident one. Moon light, in a relatively tiny amount amplify the total radiation quantitatively absorbed by the inorganic electro-catalysts dissolved. The radiant energy, related to the ElectroMagnetic and gravitational radiation [J/m²], is emitted in the surrounding empty environment on the Earth fluid surface as in the hyperspace.

Global Horizontal Irradiation (GHI) = Diffuse Horizontal Irradiance (DHI) + Direct Normal Irradiance (DNI) x cos (z)

$$SI = S_0 \left(1 + 0.034 \cos\left(2\pi \frac{n}{365.25}\right)\right)$$

Where:

SI = Solar Irradiance (or insolation)
S_0 = solar constant = 1.361 kW/m²
n = number of the day (e.g., 1st January = 1)

Photon energy SI_{Earth} = 13.8 kW/m²

$$Q_p = (h\,c)/\lambda = h\,c\,v = 7 \cdot 10^{-34} \cdot 3 \cdot 10^8 \cdot 0.5 = 3.5 \cdot 10^{-26} \text{ m}^2 \text{ kg/s}^2 = 3.5 \cdot 10^{-26} \text{ J} \approx 3(R/T)/N \approx (R\beta v)/N$$

Photon flux $\theta_q = n\,Q_p/t$ [s⁻¹]

Photon irradiance (radiant flux received) or exitance (emitted) = E_q or $M = n\,Q_p/At$ [J/m² s]

Potential of a unitary charged mass $\Delta V/1\,g = (q_p - q_e)/1g = + (R\,\beta\,v) - W = + (R\,\beta\,v) - (F\,x)$ [V]

Where:
λ = wavelength [m]
h = 6.62607004 · 10⁻³⁴ m² kg / s
n = number of photons
c = 299 792 458 m/s ≈ 10⁸ m/s
300 MHz (1 m) < v < 300 GHz (1 mm)
q_p = proton charge
q_e = electron charge
Albedo 'whiteness' = diffusive reflective solar radiation (snow 85%)
A type of actinometer, the *pyranometer*, tools for measuring the GHI (0.3 < λ < 3 nm)

Dye Sensitive Solar Cell (DSSC)

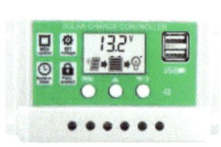

100W Solar Panel
510x280x3.0mm/20.1x11.0x0.1in

Solar Controller
133.5x70x35mm/5.26x2.76x1.4in

Figure 36 Solar cell with alligator (ΔV – 9 V, P – 3 W, I = 333 mA, 4.7 €) (left), solar panel (P = 100 W, ΔV = 18 V, I = 2 A, conversion efficiency 23 %, 58 €), LCD controller (7 €) [e.g. 1 mean house 3 kWh, 28 solar panels, each of P = 100 W => 1820 €] – Image source:
https://www.wish.com/wishlist/5de55b228d2c31f64587d23f/product/5b46c6cddc8a412e6e010f53?source=wishlist_feed&position=0&share=web,
https://www.wish.com/product/5f2001601012b468d369a8a6?from_ad=goog_shopping&_display_country_code=IT&_force_currency_code=EUR&pid=googleadwords_int&c=%7BcampaignId%7D&ad_cid=5f2001601012b468d369a8a6&ad_cc=IT&ad_lang=EN&ad_curr=EUR&ad_price=23.00&hide_login_modal=true&share=web

Dye Sensitive Solar Cells (DSSC) (Grätzel M. and B. O'Regan, 1988) made of semiconducting nano-crystalline light-scattering titanium dioxide (SnO_2 : F-doped and UV = 10 nm + Violet = 400 nm absorbing nc-TiO₂ sintered, $w_{(Sn+Ti)}$ = 12 + (100 nm - 4 μm), 10 < ø < 20)) isolated from diatom frustules, which photo-catalytic properties was discovered by Akira Fujishima in 1967, have significant potential as inexpensive photovoltaic devices. *Titanium dioxide* (TiO_2) is bio-inert extensively used in food products and as ingredients in a wide range of pharmaceutical products and cosmetics, such as sunscreens and toothpastes, characterized by an Atoxigenic *Threshold Limit Value* (TLV) = 10 mg/m^3 according the Governmental Industrial Hygienist (ACGIH(R)) (F. Grande et al., 2016). Photogenerated holes and hydroxyl radicals (-OH) were monitored over time by observing their selective reaction with probe compounds, Iodide (I^-) and terephthalic acid ($C_6H_4(CO_2H)_2$), a PolyEsTer (PET) precursor isolated from turpentine-tree *Pistacia terebinthus*, respectively. Iodide and photogenerated holes were influenced by iodide adsorption on dye TiO_2 surface, described by a Langmuir– Hinshelwood mechanism, to describe the oxidation of terephthalic acid by hydroxyl radicals. *Titanium dioxide* (TiO_2) nanoparticle photo-catalysis photoactivated by means of UV-A radiation UV-TiO_2 advanced oxidation kinetics to yield the degradation of oxalic acid, reference metabolite, in a dispersed-phase photo-reactor. $5 < [TiO_2] < 400$ mg/mL initial $0.9 < $ [oxalic acid] $ < 32$ mg/L, $100 < $ lamp irradiance $ < 10^4$ W/m², $0 < $ background fluid absorbance $ < 30$ m^{-1}, $1 < $ reactor size $ < 4$, $0 < $ lamp orientation $ < 360°$, and $2.5 < $ flow rate $ < 10$ m³/ h). The analysis revealed that an optimum in oxalic acid degradation is observed when the $20 < TiO_2 < 40$ mg/L (D. Santoro et al., 2017).

DSSC made using UO_2 instead of TiO_2, would be six folds heavier and almost twice electrically gaining, making this chemical compound more than a nuclear fuel, but an electron acceptor in a transduction chain from solar, through nuclear to electrical energy. It has to be managed with high safety care as it can easily bioaccumulate and induce cytotoxicity in the heart, brain, bone, stomach, gut liver, kidney and gonads. Traditional silicon-based solar cell offers about 35 mA/cm², whereas current DSSCs offer about 20 mA/cm², but DSSC works only at very low-light condition and furthermore are robust, stable, and due to the light weight easily roof transporting, hence, suitable for portable collectors in cold countries. The quantum efficiency of a DSSC depends on four energy levels of the component: the excited state (approximately LUMO) and the ground state (HOMO) of the photo-sensitizer, the Fermi level of the TiO_2 electrode and the redox potential of the mediator (I^-/I_3^-) in the electrolyte (Hara Kajiro, 2005).

A series of zinc tetraphenyl porphyrin photosensitizers furnished with three different anchoring groups [phenyl-phosphonate] = 156.08 g/mol prepared using *'click methodology'*, adsorption behaviour was modelled using the *'Langmuir isotherm model'* finding that affinity for Nickel Oxide [NiO] = $6.65 \cdot 10^4$ M interphase and the fastest rate of adsorption A = $2.46 \cdot 10^7$ cm²/mol · min (https://doi.org/10.1098/rsta.2018.0338). Photogenerated holes and hydroxyl radicals (OH^-) were monitored over time by observing their selective reaction with probe compounds, iodide (I^-) and terephthalic acid ($C_6H_4(CO_2H)_2$), a *PolyEserTeraphtalate* (PET) precursor isolated from turpentine-tree *Pistacia terebinthus*, respectively. Iodide and photogenerated holes were influenced by iodide adsorption on titanium dioxide (TiO_2) surface, described by a *'Langmuir– Hinshelwood mechanism'*, to describe the oxidation of terephthalic acid by hydroxyl radicals. TiO_2 nanoparticle photo-catalysis photoactivated by means of UV-A radiation UV-TiO_2 advanced oxidation kinetics to yield the degradation of the reference metabolite oxalic acid in a dispersed-phase photo-reactor. $5 < [TiO_2] < 400$ mg/L), $1 < $ [oxalic acid] $ < 32$ mg/L, lamp irradiance $100 < I = E/A < 10,000$ W/m², background fluid absorbance $0 < A < 30$ m^{-1}, reactor size ($1/4 < $ relative scaling factor $ < 4$, lamp orientation ($0 < \alpha < 360°$) and $2.5 < $ flowrate $ < 10$ m³/h. The analysis revealed that an optimum in oxalic acid degradation is observed when the $[TiO_2] \approx 30$ mg/L range. (D. Santoro et al., 2017).

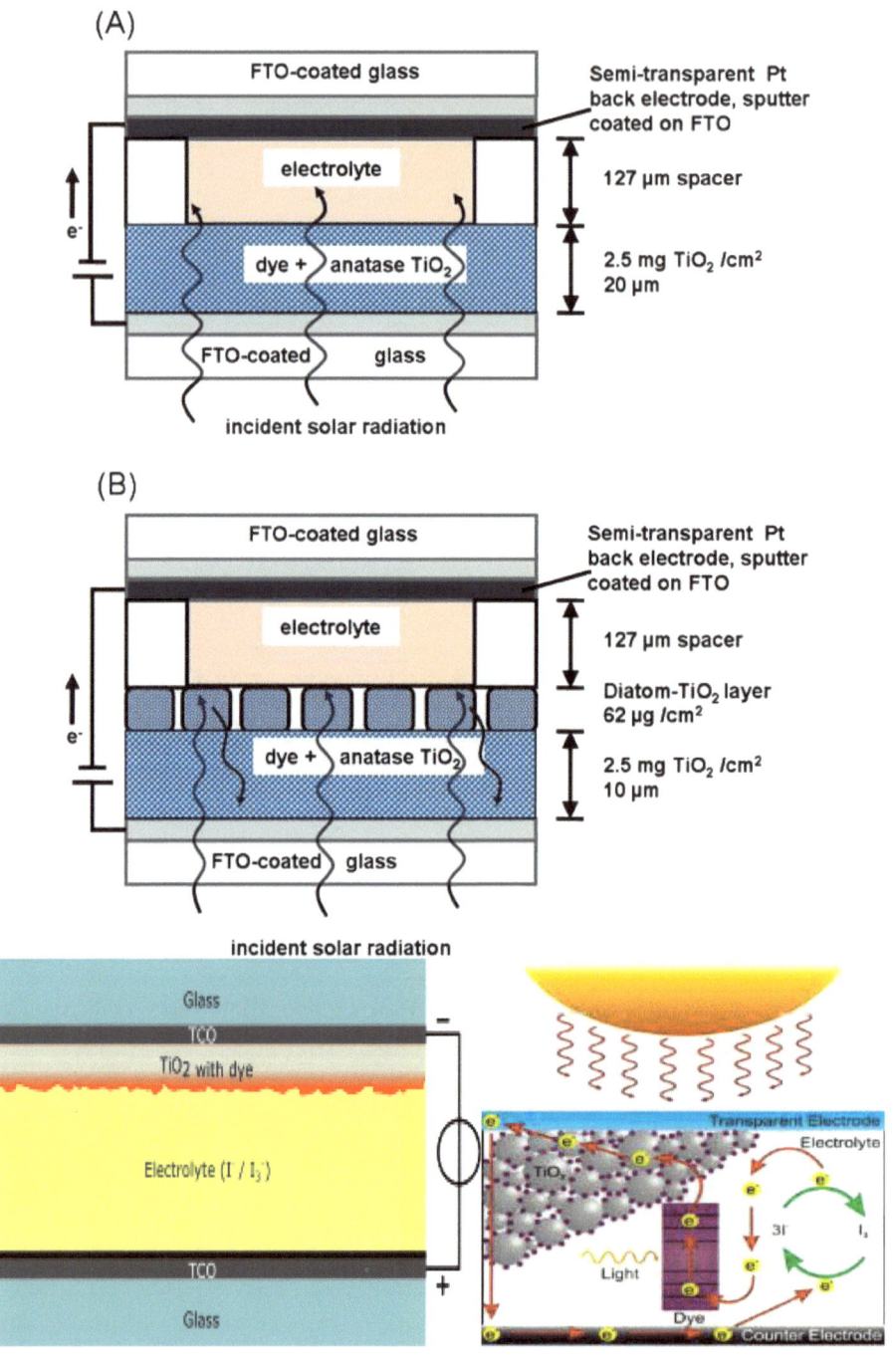

'Wormhole *in vitro*' © Copyright Antonio Silvestro, 2020

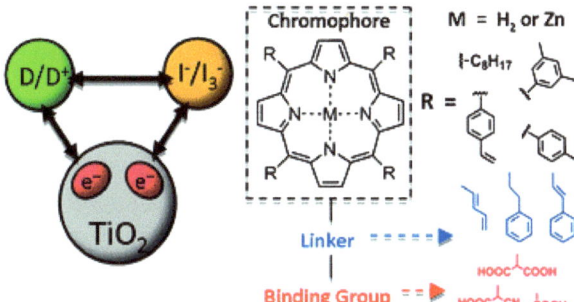

Figure 37 (A) Traditional DSSC, (B) DSSC with enhanced photonic capture (1, 2, 3). Porphyrins for DSSC: new insights into efficiency-determining electron transfer steps (4) - Image source: https://pubs.rsc.org/en/content/articlelanding/2012/cc/c2cc30677h#!divAbstract

The photon absorption phase of the PVE is eye-visible on the ocean surface as a mesmerizing optical reflection entrapped in the water surface, phenomena happening when the bright *Helios* let see its wonderful reflex to the sentient mortals capable of receiving its photon-signalization via **pigments** like the ocular *Mercury rhodopsin* (reddish) or photo-synthetic pigments like the *phycoerythrin* (reddish), *Jupiter bacteriorhodopsin* (pinkish) and *chlorophylls* (a, b, c, d) (greenish), *fucoxanthin* (deep green, brownish), *Saturn phycocyanin* (bluish), *Venus carotenoids* (yellowish *xanthophylls* containing -CHO_2 and orangish annatto bixinine and norbixineand reddish lycopene *carotens*–CH), *Neptune anthocyanins* (violetish). Dyes types: $_{44}$Ru (4,4'-dicarboxy-2,2'-bipyridine)$_2$(NCS)$_2$, Ruthenium-poly-pyridine, Perovskite ($CH_3NH_3PbI_3$), Haemoglobin, Phyco-phyto-pigments Mg like-chlorophyll.

Figure 38 [Ru(4,4',4"-(COOH)$_3$-terpy) (NCS)$_3$], effective at low frequency (low solar energy), 80 % photo-electric conversion. 1-Ethyl-3 *MethyLimidazoliumtetroCyanoBorate* [EMIB(CN)$_4$] which is extremely light- and temperature-stable, Copper-Di Selenium [Cu(In,GA)Se$_2$] which offers higher conversion efficiencies.

DSSC [ΔV_{module} 36 cells = 18 V, ΔV_{cell} = 1 V, I_{cell} = 500 mA, P = V x I => 100 W, 18 x 5.6 (5 €/cell)]:

1. Weight on a scale the bioplastic, genitak, textolite, temper PolyStyrene (PS) glass or PolyVynilCloride (PVC) supporting background (3 x 70 x 70 mm);
0. Light absorption by dye molecules;
1. Electron transfer to titanium dioxide (TiO$_2$);

2. Brush catalysing ethanol (C_2H_5OH) on 2 busbar columns;
3. 10 fingers rows done of anodic nickel chromium (NiCr) wire ($0.1 < ø < 2.5$ mm) alternated to cathodic copper (Cu) wires ($ø = 0.2$ mm);
4. Weld busbars and fingers with a Hephaestus solder iron ($P = 70$ W);
5. Flux pen for preventing the solder beading and helps the solder flowing cleanly onto the soldered parts;
6. Brush cyanoacrylate superglue or slow coating transparent crystalline pasting epoxy resin ($m_1 = 100$ g), and gently mix with a toothpick or stirring glass rod the sticky yellowish hardener ($m_2 = 10$ g; 1:10) between Si-wafer and support for about $t ≈ 8$ min, encapsulating the solar cell;
7. Set it for $t = 45$ min till all the gas will evaporate piercing eventual air bubbles with a needle;
8. Solidify the gluing cover for $t = 24$ h at room temperature or increasing the drying speed using a hot incandescent lamp;
9. Percolation to the anode's surface turning the product upside-down.

White Hole cathode (+) working electrode:

1) Making the paste: diluted Nitric Acid ($HNO_{3(g)}$) x TiO_2;
2) Testing conductivity: $R = 0$ Ω;
3) Stick with taper the glass on a paper foil;
4) Place the paste on the top of the glass;
5) Sitting rod to press and push the paste across the glass and leave it dry for $t = 5$ min;
6) Remove the tape leaving the square of paste dried on the glass;
7) Dying the cell: blackberry juice (*Rubus fruticosus*) on top;
8) Wash with isopropanol the cell into a beaker;
9) Rinsing and drying;
10) Current to the cathode in the external circuit.

Black Hole anode (-) counter electrode:

1) Graphite catalyst colouring a glass with a pencil;
2) Clip the two electrodes together face by face making sure the off-site edges;
3) 2, 3 drops of redox electrolyte I^-/I^{3-} and KI;
4) Electron transfer to the electrolyte;
5) Charge transportation through the electrolyte;
6) Dye regeneration;
7) Aluminium (Al) frame and L-shaped corner junctions sealed with silicon.

Test the DSSC with a multi-metre with crocodile clumps measuring the physical quantities values (e.g. *Power Bank*, solar charging: $ΔV = 5$ V, $100 < I < 300$ mAh, Input 5 V/A, $14 <$ Price < 17 €).

PhotoVoltaics (PV) panels

PhotoVoltaic (PV) quality standards according the *Iustum*, precisely, to respect International Electro-Technical Commission (IEC) 61215, International Organization for Standardization (ISO) 9488, UL 1703-41-2703, CE, Electrical Safety Tester (EST) 460-22V-22H-110. Energy conservations measures

- Feed-In Tariff (FIT) incentive scheme. Electricity law 46/90.

In Italy, point of sending and delivery (PDR and POD) are used in the distribution of energy such as light and gas. Electricity Bill Calculator ENEL gas and light 0.05 €/KWh = 180 €/month - https://www.rapidtables.com/calc/electric/electricity-calculator.html e.g. 1 laptop t = 16 h, P = 65 W + cooking range t = 2 h P = 400 W, fridge t = 24 h P = 188 W, coffee pot t = 1 h P = 400 W, fun t = 16 h P = 40 W + 2 Compact Fluorescent Lamp (CFL) t = 3h P = 20 W). Standard meter cubic to KWh
(https://www.google.com/search?q=conversion+smc+to+kwh&rlz=1C1GCEA_enIT854IT855&oq=conversion+smc+to+kwh&aqs=chrome..69i57j0l2.6845j0j7&sourceid=chrome&ie=UTF-8) [1 KWh = 3.6 MJ]. E = 3 kWh = > P = 3 kJs/3600 s = 0.8 J/s = 0.8 W = 0.2 cal/s. ENEL 86.4 kcal/day vs Human diet 2 k cal/day. As in a mean family there are four people, mother, father and a couple of babies (boy and girl), electro domestics needs about 10-folds the daily human consumption.

D. Lgs. 222/2016 installation photovoltaic panels, implant from renewable sources.
D. Lgs n. 387/2003, art.1 thresholds authorization needed.

$$\text{PhotoVoltaic} > 20 \text{ kW}$$
$$\text{Eolic} > 60 \text{ kW}$$
$$\text{Hydro-electric and geothermic} > 100 \text{ kW}$$
$$\text{Biomass} > 200 \text{ kW}$$
$$\text{Biogas} > 250 \text{ kW}$$

Comma 5 Under the aforementioned threshold Signal of Initial Activity (SCIA) (http://www.comune.casoria.na.it/scia2 modules updates)
(http://www.ingegneri.info/news/edilizia/scia-edilizia-al-comune-di-napoli-come-presentare-la-modulistica/)

D. Lgs. 28/2011, art. 6, c. 11 Communication to Region and Province of PV from renewable establishment till P = 50 kW.
D. Lgs. 28/2011, art. 7 Communication for Thermic implant that doesn't modify the building shape.
D. Lgs. 28/2011, art. 7-bis Communication Micro-generation of electricity.
D. Lgs. 28/2011 art. 8 Authorization/Silence assent for bio-methanoids.

Luminosity power rating of panels	100 W
Charging	5 h
Current/h	100 Ah
Number Batteries	2
Batteries voltage	12 V
Number Solar Panels	3
Surface area covered	0.429 m^2
Ratio Ch control	25 A
Inverter power	713 W
Max Solar Insulation	6 kWh/m^2/day in July (Gaussian Curve)
Peak Sun hours based on half-sine models	5.5 h/d
Max Irradiance	1 kW/m^2 at noon
Solar panel inclination tilt, tracking and shading	Winter 17 < α < 64° Summer

'Wormhole *in vitro*' © Copyright Antonio Silvestro, 2020

Orientation	(S)
Rayleigh, Lambert Beer, Absorbance PAR, Radiation Spectra A. H. Becquerel	
http://www.solaronix.com/technology/assembly/	

Solar Calculator software for efficiency and power needed assessment. *SketchUp* (https://www.sketchup.com/) and *Skelion* plug in (http://www.skelion.com/index.htm?v1.0.1). Solar panel model (https://3dwarehouse.sketchup.com/search/?q=solar%20panel&hl=en or

https://3dwarehouse.sketchup.com/search/?q=solarpanel&hl=en).

System Efficiency 70 %, Offline Usage - %, Depth of Discharge - %, 10274 Wh used in 1 day. 30 % E_q of the incident solar power $P = 166 \cdot 10^{15}$ W (= 166 PetaW) is reflected into space, 19 % is absorbed by the clouds, 50 %: P = 85 PW available for terrestrial energy harvesting, of which 88 % is utilized for primary production due to photosynthesis (40 % E_q), almost five-fold the human power consumption in 2010, in other words, *Homo sapiens* utilize just the 8 % of the total solar radiation.

Crystal Silicon Glass (CSG) made via *purifying Si* from sand (S + C at T = 2000 °C -> Si raw + HCl heating -> $Si_{(g)}$) followed by PECVD polymerizing silicon at T = 500 °C on borosilicate (poly c-Si ø = 1.5 µm, r_{Si} = 11 Å, A = 0.5 cm^2) with cell efficiency = 23 % and module efficiency = 13 % for obtaining panel costing 50 €/m^2 [40 €/panel (0.4 €/W)]. *Third-generator* photo-voltaic cells are able to overcome *'Shockley-Queisser'* limit of 35 % power efficiency for bandgap, *Czochralski* process for mono-crystalline Si cells making, or multi-layer tandem made of Si and *Gallium Arsenide* (GaAs), *Selenide* (GaSe) or *Cadmium Telluride* (CdTe), up to 62 % efficiency and electricity I = 850 mV. **Photon cavities concentrators** or light-receiving semiconductors as nano-cylinders (e.g., Al, Ag price$_{Ag}$ = 43.52 € or Au that absorbs UV and scatters vis-IR) and filtering lenses (SiO$_2$ + BO -> B$_2$O$_3$ low melting borosilicate) and canalizing mirrors (Al or Ag coating SiO$_2$). **PhotoVoltaic module** (A = 1 x 1 m, 40 < P/module < 180 W) assembly of 30 square cells/module (0.5 < V/cell < 0.6), in which two electrodes (ΔV_a = - 0.5 V, ΔV_c = + 1 V) thicker parallel metallic busbars ribbons (Cu 1.72 · 10^{-8} Ωm, price$_{Cu}$ = 0.0091 €/g or Ag 1.59 × 10^{-8} Ωm, price $_{Ag}$ = 0.51 €/g) and many thin fingers bars perpendicular to them forming a distribution electron network laying on a wafer background done of organic compounds, amorphous or Mono-/Multi-crystalline Silicon (c-Si wafers (or bulks) of thick up to w = 200 µm). PV combinations: (Cu, Ag or Au) x (Al, Ga, or In) x (S, Se or Te) – e.g. cheapest CuAlS. Inverter conductive **wires** (Cu, Ag or transition metals group d) in parallels for converting electric Direct Current (DC) into Alternating Current (AC), while, in series for voltage difference (ΔV) supply. Ceramic-alloy *Indium Tin Oxide* ($_{49}$In$_{50}$Sn$_{16}$O), *Copper Indium Gallium Selenide* CIGS ($_{29}$Cu $_{49}$In $_{31}$Ga $_{34}$Se), or Ir-Si aerographene for insulating the electrodes.

Autotrophic inoculum for White Hole anode (+)

The Bubble Universe originary event (O), or succession of it would have been done of an autocatalytic series of spontaneous interstellar glycolaldehyde formation for RiboNucleic Acid (RNA) replication and organic macromolecule assembly according the second law of thermodynamics, followed by the separation and isolation of the self with self not due to the emergence of membranes enclosing LUCA within it, through which inorganic elements would displace from and to the primordial environment (chemiosmosis) ruling the **abiogenesis**.

The Black Hole generated after the explosion of the Supernova that made the Solar System (SS) planet be, had had its counterpart the White Hole at the otherside of a brane from where water (H₂O) should have been arrived. The instantaneous data teleportation from a cone to another between the holes of the wormhole, what would happen also in human subtle bodies transferring quantum-bits from an anatomic system to another via the circulatory systemically. Certainly, not hand by hand in a circle would be accomplished the resonance capable to teleport unidirectionally information without damage, as the first would lead to the over-whelming of the centre, while in the last it could flow from A to B unalerted bringing within the spherical singularity microbes on the linage of LUCA.

The cycling chemical origin of life on Earth described by *Miller-Urey* (1952) can be revisited in an **Ab Initio Molecular Dynamic (AIMD) Solar System (SS)** scale, in which Jupiter Singularity is proposed as a glass sphere in which water (H₂O), methane (CH₄), ammonia (NH₃), hydrogen (H₂) are under Neptune vacuum and ionized into plasma electricity flowing into sparking Uranus electrodes, descending in a the Saturn cooling condenser passing through the transitioning Mars gate, before entering in the heated underworld of Venus, the death of the mortals on Earth evaporating their common solvent in the generative Mercury solute methane (CH₄), ammonia (NH₃) reagents of the Singularity in which 20 proteogenic amino acids were spontaneously formed. Certainly, adding microorganism descendant from LUCA in this edited version could clarify the biochemistry behind evolution in the planetary system in which *Homo sapiens*, and unofficial most evolute *Homo atm* are living in. The borosilicate glass that insulates the simulation, would be in the inspiring experiment geoidal atmosphere, while, the innovative the interstellar medium and solar wind pressure-balanced bow shock H₂-boundary: the ellipsoidal heliopause. After choosing which *Eubacteria*, *Archea* or *Eukaria* place in the Singularity, could have been added to it also the obligate parasite viruses for understanding their astronomical origin. Would this entity have been evolved from a parallel Universe resonating with the artificially proposed? Two copies identical one at each side of the bridge, you and I, my ego the instant before deciding the transitioning to a different state of mind widening the soul that make the invoker of the absolute an empty *Homo atm*.

Kohn–Sham (KS) formulation of density functional theory of the AIMD, wherein the total energy is expressed as a functional of n mutually orthonormal single-particle electron orbitals $\psi_i(\mathbf{r})$, $i = 1,..., n$. (https://doi.org/10.1073/pnas.0500193102). Is hypnotizable that the originary density would be somewhat proportional to the density of the human heart: ρ_{heart} = 1.055 g/mL.

Is plausible that **Last Universal Common Ancestral (LUCA)** would have an inductive toroidal magnetic field wormhole-like, while, the humans would have added due to environmental and genome evolution other five pair of perpendicular cones parallel among them corresponding to the **state of matter** (*Mercury* solid particles, *Venus* liquid droplets, *Jupiter* plasma sparks, *Saturn* condensate lenses and *Uranus* gas bubbles), but the cumbersome *Singularity* the homologue originary cell of all the organisms existing in the Bubble Universe, that with its roto-translation would generate the conical spiral electricity of the Kundalini emitting light via Albert Einstein photoelectric phenomena. Two opposite in charged boson photon in entanglement emitted from the artificial WormHole super-conductor with zero electrical resistance and expelling the magnetic field from its inner self exited, would be mixed in the heterodyne of the CWB generator connected to it would displace along one resulting direction oscillating between Black cathode (-) and White anode (+) continuously into Alternating Current (AC).

Strong, weak, gravity and EM the main forces, this lasts the one that challenge with the penultimate, while the first two act similarly, but on a different scale. For the teleportation experiment would have be chosen one only microorganism generating electric and inducing magnetic field among the sentient being after the LUCA according 16S rRNa analysis is the simplest sentient organism lived about 3.7 BYA from which *Eukarya, Prokarya* and *Archea* descend silencing everything is currently express in the community of living form existing in the Bubble Universe. Whenever it would be found, a genetic engineering recombinant of electro- and magneto-genic microbes could be created in lab using among the model organisms already existing.

In this biological version the wormhole in which life would be teleported over the fringe between two holes, a synthetic dye, LUCA descendant microorganism or their isolated pigments and cofactor like the Mg of the chlorophylls would be placed in the Singularity. *Archaea* able to accomplish photosynthesis actively catching Sun light, injected with syringe into the generative spheroid medium for feeding the feedstock that would remake the anaerobe, autotrophic CO_2 fixating via the acetyl–CoA (or Wood–Ljungdahl) pathway, H_2 ferredoxin reducer, LUCA at the beginning of the Bubble Universe where just *Hades* were ruling the dark matter of the hell, before the blooming of the paradisiac life that we all diversely live or deserve to live through. Well, tears in the underworld a part, remember that the engineer of the present would choose among many photovoltaic cells sandwich, variable in the layers assembly, depending on the yield requirement, and find the ongoing innovation in the capability not just to absorb, but also directly emit light in the past generation let them rebirth continuously into our bouncing heart from future prospective and their memories, without need for the psychopomp Charon carrying the remnant electrons into an external circuit where they can rest at the Albert Einstein energy, but switched along the wall of life and death, in a way inducing 'Hall effect' where just *Chronus* and *Rhea* would decide which electron-soul would pass through the membrane of the depletion hypersurface ($P_{max} = \Delta V \cdot I = 0.5 \cdot 7$ mA/cm^2 = 3.5 mW) testable with solid *Polaron* (P^+) quasi-particle verifying the polarization between positive P and negative N bordering junction (1 < charge carriers' diffusion < 10 nm), e$^-$- hole pairs hosted into the *Jupiter* Singularity, producing a driving force (F), potential difference (ΔV) and electric current (I) via *Juno* superconductive globular coil analogously to PV fingers and busbars, let them feel the eternal silent dying in the redox shuttled dye that would photo-electrically emit quanta in bright nirvanic violetish light of continuously rebirth bouncing between Black and White Hole. The synthetic wormhole could be P-doped adding inorganic *Phosphorous* ($_{15}P$) mimic the organic in the backbone of the DNA, injected via a syringe through the insulator generating an electric field (E), exciting the lepton electrons from $3s^2$ to $3p$ (s = spherical, p = prolate ellipsoidal orbitals) catching boson photons, for later displacing from N (-) into P (+) region, where the of the metalloid boron ($_5B$) injected, at room temperature (T = 25 °C), into the charge separative integrated electrical connection on glass monolithic gate, would act as cavity for hosting them [3rd group: 3 valences e$^-$ in the (second) outer orbital, the elemental cavity hole where other 5 electrons can flow] ionizing five protons (5 H$^+$) and five electrons (5 e$^-$), allowing the 20 fold decrement of the activation energy, actually, from 1.12 to 0.05 V. The jam of charged particles would push the generative spheroids from one side to another facilitating the enzymatic dye photons absorption that would trigger the electron transport, polarized according the electro-potential of the outer Ultra High Vacuum (UHV) medium where just the cuddling Cosmic Wave Background (CWB), mother of all the corpuscles in us generating diatomic hydrogen (H$^+$ + e$^-$ -> H$_2$), through us and forward our own sentient being, humans capable to feel infrasound, to hear audio frequency, to get tanned or even burned by ultraviolet. And do not say that you have never heard the voice of your followers hammering the piezoelectric of your heart. *Superfluid helium* (He) could substitute nitrogen in the vacuum bilayer surrounding the border of the architecture Universe bubble of the Big Bang 1.3 million years ago, that certainly, didn't needed an insulating nickel shell as drawn by Jordi Prat-Camps Carles Navau and Alvaro Sanchez liming the lemniscate whole of the existence in a domain of finite calculi, as the mu-metal cease to exhibit spontaneous magnetization above the Marie Curie temperature, where thermal motion-entropy, overcome the ferromagnetic tendency of dipoles like the Singularity and the artificial Einstein-Rosen bridge itself to align with an external reorienting magnetic field (e.g. T_{Curie} = EuO 69 K = - 204.15 °C, Ni 627 K = 354 °C, Fe 1043 K = 770 °C), exhibiting paramagnetism, aligning their unpaired dipoles in parallel to an external EM field like the one induced by relict CWB generator triggering the sonoluminescence of the Universe Bubble that emitting light would vanish showing the ceiled wormhole within it.

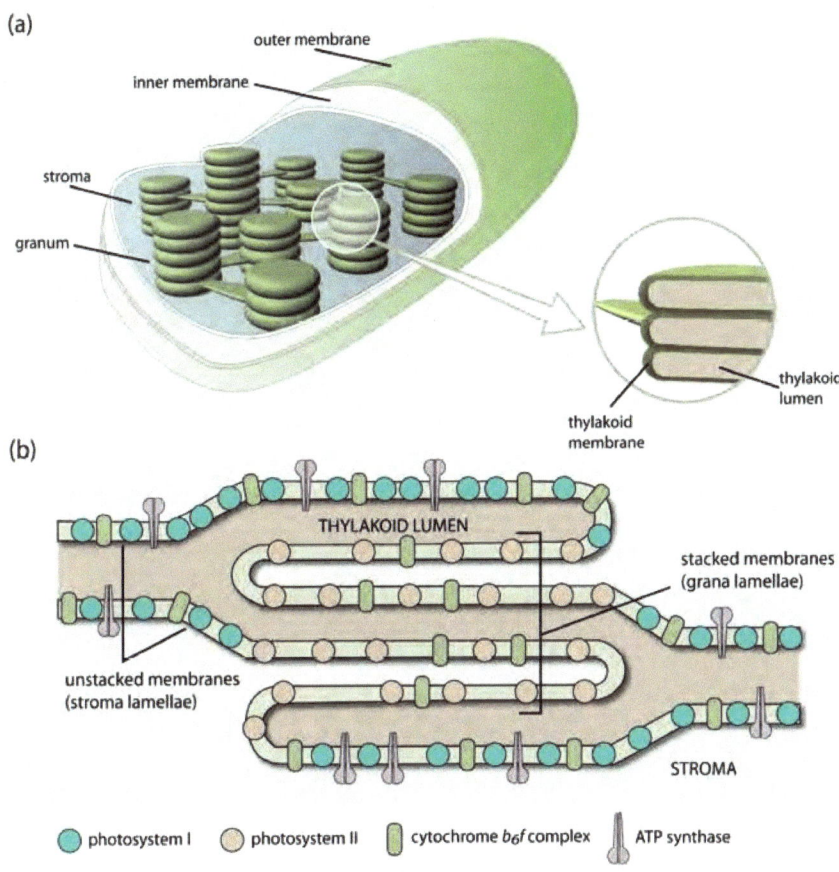

Figure 39 *Eukarya* organelles chloroplasts and *Prokarya Cyanobacterias* (e.g., *Arhtospira fusiformis*) in which the electrodes disks interleaved biconvex thylakoids lamellae separated by the thylakoid electrolyte lumen in which is dissolved the copper (Cu)-protein plastocyanin form the so-called grana immersed in their stroma, are the biological counterpart of the voltaic piles.

The **photosynthesis** phenomena is a the three stages process: sunlight harvesting artificially reproducible by with wide spectra incandescent lamps, and energy storage into Adenosine TriPhosphate (ATP) bonds and reducing power Ferredoxin and Nicotine Adenine DiNucleotide Phosphate (NADPH), Calvin Cycle naturally accomplished by aerobe plants, algae, *Cyanobacteria* reducing and fixing the Carbon Dioxide (CO_2) into carbohydrates [$(C(H_2O)$)], by anaerobia sulphuric bacteria using Hydrogen Sulphate (HS) and thiosulfate ($S_2O_3^{2-}$) or also non-sulphur microorganisms using fats and amino acids as electron donors. The tree photosynthetic stages in the 'Wormhole *in vitro*' would be remake in the Black Hole, hypersurface and White Hole, respectively. In higher plants, *Magnolipsidsa* (or *Angiospermata*), an average of 2500 Chlorophylls (Chls) are needed in the light harvesting antennae complex to absorb light and transfer it in circa t = 1 h to the reaction centre where four photons activate the hydrolysis of one molecule of water (H_2O) releasing an oxygen (O_2). As the primary production is based on photo-synthetically active organisms that can harvest sunlight, the purest source of energy of our system, with their pigment antennae in the photosystems, leaves quantum of light absorbed jump among them till the center of reaction releasing four Hydrogens (H^+), stored as short-term energy as Nicotinamide Adenine Dinucleotide Hydrogenated (NADH) or Adenine TriPhosphate (ATP) via ATPases, and activating an electrochemical signaling through proteinaceous membrane complexes using reducing power for fixing CO_2, through the Calvin cycle, accumulating long-term energy as sugars photo-assimilates. The most conserved enzyme Ribulose 1,5-bis-phosphate carboxylase/oxygenase, that take part in C_3 cycle, that can be fixed by cryo-electron high-resolution imaging showing a larger subunit coded by the genome and a smaller expressed by the plastome that came from nitrogen-fixing oxyphototrophic prokaryotes *Cyanobacteria*. RuBisCo is a mere evidence of **symbiogenesis**, basement for the co-evolution, hence, for the human evolution itself, just thinking about the human microbiome four-fold richer than our own. 'Life did not take over the globe by combat, but by networking' (Dorion Sagan and Linn Margulis, 1967), where life is not just living to survive under a selective pressure, but to flourish helping each other, where there is no power in the hands of a monopolistic being.

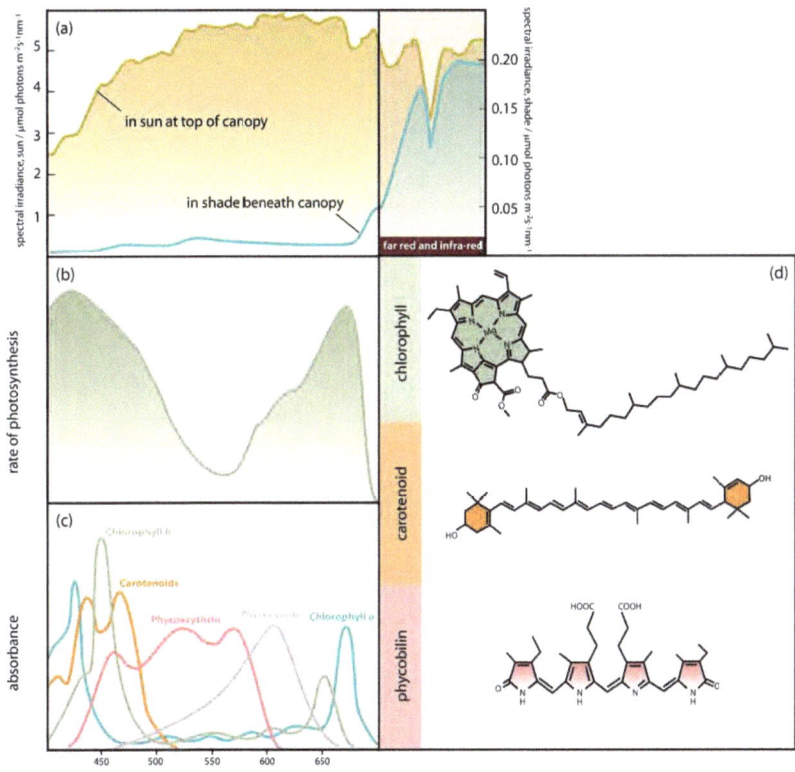

Figure 40 Pigments spectral absorption and molecular formula. P.S. In the Far-Red and Infra-Red (700 < λ < 1000 nm) absorption are independent from shade. Infra range among the knots of the chlorophyll waves, otherwise, corpuscle peaks in the PAR range, Chl a = 125 nm, Chl b = 260 nm – Image source: Thina Mirkovic et al., 2016.

Magnesium ($_{12}$Mg), alkaline Earth metal with the lowest melting (T = 650 °C) and boiling point (T = 1090 °C) of electrical conductivity equal to σ = 44 S/m (at T = 20 °C) shiny grey brittle solid form when a star age itself adding three α particles (2 H$^+$ and 2 neutrons each) to a $_6$C nucleus, is the third most abundant element in seawater after sodium ($_{11}$Na) and chlorine ($_{17}$Cl), hexagonal crystal characterized by an invariable oxidation state = + 2 when bonding other elements like oxygen ($_{16}$O), forming (2MgO) with which it exothermally ignites emitting bright white light in the UV. It has three stable isotopes: ^{24}Mg (79 %), ^{25}Mg, ^{26}Mg, while, ^{28}Mg is radioactive with a short half-life (t $_{½}$ = 21 h) can be produced in laboratories by electrolysis of saturated hydro-solution, but it has to be managed with care, thermoplastic UV-filtering eye glasses and insulating gloves, as its powder is highly flammable (T = 3100 °C, height = 3 cm), temperature value that decrease 6-fold acquiring a 3D ribbon configuration (rectangular band knotted showing from 1 to n loops) as in the Grignard reagents (R-Mg-X). Hydrogen (H$_2$) could be produced controlling the hydro-solution flammability insulating the ellipsoidal (or just spherical) system with silica powder on a metallic scaffold

'Wormhole *in vitro*' © Copyright Antonio Silvestro, 2020

characterized by a higher melting point than magnesium, melting with the same H_2 hot steams, in other words, the enclosing walls are created in cycle on the destructive burning feedback. Disposable Mg primary batteries and rechargeable accumulators are on market and in prototyping status, respectively. This element is present in the adult human body (24 g) mainly in the skeletal (60 %), skeletal muscle (0.8 %) displacing as $[Mg^{2+}]_{serum}$ = 0.85 mM = 2.1 mEq/L – *Mulhadhara* chakra, being in its cation form antagonist to Ca^{2+} an essential secondary cofactor enzyme-binding like the ATP synthase, bond to the porphyritic tetra-ring of the haemoglobin flowing through circulatory and digestive systems, released into the stomach where it bonds Tricarboxylic acids enclosed in the Krebs cycle, moving in the gall bladder merging with the malate involved in synthesis and β-oxidation of fat acids and through the cilia in the intestinal lumen in the opposite direction of calcium generating electrical potential through the plasmalemma. Last, but not least magnesium in the nervous system, linked mainly to GABAergic, glutamatergic, dopaminergic, adrenergic, and serotonergic. The accomplish the overview the serotonin neurological pathways are photo-electrically relocating via the only endocrine gland, not related to the *Vishuddha,* but to the higher *Ajna* chakra, the pineal gland and the immune system via *CerebroSpinal Fluid* (CSF) wave resonance within, can be mathematically approximated to a set of corpuscles tied and/or threaded, a string placed in the spinal cord in turn in the backbone cavity which stretchable fibrocartilages disks between each vertebra thicken themselves till reaching their maximal height at an average of 22 years old resonating within the *Betelgeuse* of *Orion* constellation.

Long-lasting (t = 80 h) **Photo-Electrochemical Cells** (PECs) produce current and/or hydrogen via the hydrolysing of water (H_2O) into hydrogen ($V_{H2(g)}$ = 150 mL: $E_{H2(g)}$ = 2 kJ) and oxygen (O_2) (4 e⁻ + 4 H_2O -> 2 H_2 + 4 OH^-) ('artificial photosynthesis') photolysis catalysed by semi-conductors [e.g. Niobium ($_{41}Nb$), Titanium ($_{22}Ti$), Platinum ($_{78}Pt$)] or metal complexes dissolved (e.g. GaInPN).

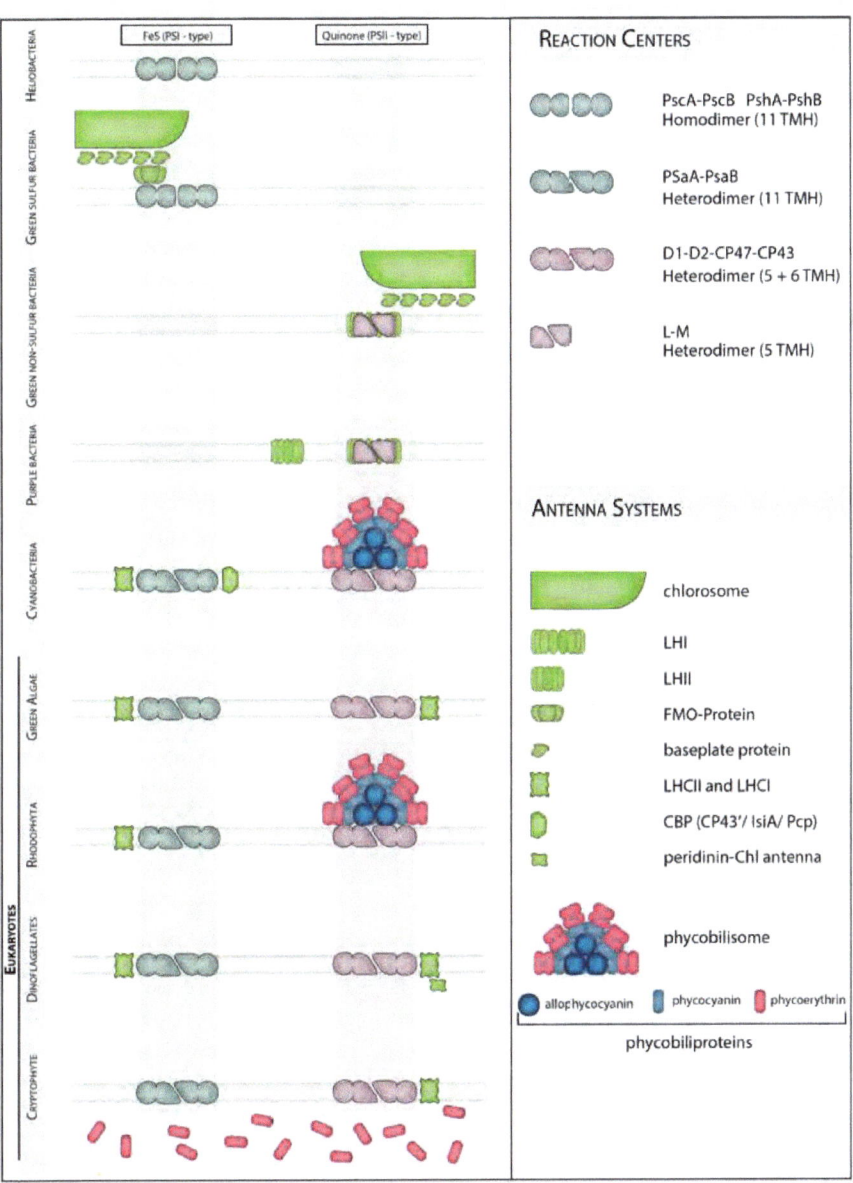

Figure 41 Photosystems.

Biological PhotoVoltaics (BPV) utilize photo-autotropic microorganisms such as Phylum *Cyanobacteria* (3 GYA) orders *Gloeobacterales, Synechococcales, Spirulinales, Chroococcales, Pleurocapsales, Oscillatoriales, Chroococcidiopsidales* and *Nostocales* may be used as biofuel living feedstock, for generating electricity injecting them into semiconductive-insulating ITO-glass windows active photosynthetic panels, for boosting the performance of technological fuel cells and even Integrated Circuits (IC) for epidermal human biohacking patches. When used in Microbial Fuel Cells (MECs) are inoculated in the anode chamber (-), managed carefully as they can express cyanotoxins (neurotoxins, hepatotoxins and dermatoxins), for catching sunlight and transducing it into electricity. The mitotic fuse of the zygote in division is shaping according the magnetic field line of the biological cell in cytokinesis as the planet Earth geomagnetic field meridian and the nadi passing though the Anahata chakra and Manas Sharir hear subtle body. If a chromosome done of two chromatids place itself on the equator as in anaphase and the centrioles are cutted away the fuse may assume a double Einstein-Dirac cone configuration and the cell is electromagnetic shieled in a Mu-ferromagnetic (Ni/Fe) cylinder assuming the shape of a spacetime clepsydra that would you bring back to the origin when the creation was orientated by the magnetic field in the hypersurface of the present where the electrodes chromatids are laying, one at the anode (-) and another at the cathode (+), in the cellular electrolyte. Salacia sodium chloride (NaCl) would dissolve into Na^+ and Cl^- diffusing through a Nafion Proton Exchange Membrane (PEM) to the antipodal electrodes at North (-) and South (+), respectively. The photoautotrophic *Cyanobacterium* (e.g. *Synechocystis* sp. PCC 6803), close to LUCA, in the oceanic solution of your *Shahashrara chakra - Nirvana sharir* would bring to rebirth from the past of your memories the highest energy range of the Photosynthetic Active Radiation (PAR), the violet light that delight the Sumerian God Utu, the Egyptian Amun-Ra, the Sanskrit Surya, the Greek Helios, the Latin Sol and the Japanese Amaterasu Kami as a dot inscribed in a circle as the Monad and the astronomical symbol of the Sun itself, generating about $1 < P/A < 610$ mW/m^2 using *Synechococcus* sp. BDU 140432. On the other side, at the cathode (+), heterotrophic waste digester *Eubacteria* like *Shewanella oinedensis* could increase the potential difference between the two-chamber breathing in the acid medium triggered by the protons released by the *Cyanobacteria Arthrospira fusiformis* and passing the fridge to the Hell where Ereshkigal/Yama/Hades/Pluto/Shinigami rule. In the rebirth cell (-), there are repeated photosynthetic cycle, while, in the death cell (+) happen the regeneration of molecule bringing life – H_2O via saprophyte coprophage microorganism breaking down shit. The Indium Titanium Oxide-PolyEthylene Terephthalate (ITO-PET) anode (-) cone base would have to be transparent, while, the cathode (+) lid opalescent, sealing respectively alkaline (e.g., calcium carbonate $CaCO_3$) aerobic and an acid (e.g., hydrogen chloride HCl) anaerobic halves. Alternatively, the cathodic chamber may be governed by Titan Selene/Luna-Olympian Artemis/Diana/Nanna/Chandra/Tsukuyomi, but it in this way it cannot be linked to the hole generation of the Universe but just to the life on Earth. The moon, fifth of the satellites for extension and largest of any of the dwarf planets, at a distance form Earth d = 385 Mkm, rotating around it in t = 27 days, is superficially composed about by 45 % silica (SiO_2), 21 % alumina (Al_2O_3), 12 % lime (CaO), 14 % iron (II) oxide (FeO), 7 % magnesia (MgO), 3 % titanium dioxide (TiO_2), 0.5 % sodium oxide (Na_2O), in which dark craters bottoms temperature colder even than on Pluto has been recorded at the winter solstice T = - 247 °C.

Bio-anode (-) inoculum of oxygenic aerobic photo-trophic cyanobacterium *Synechocystis spp.* PCC6803 (freshwater) or *Cyclotella cryptica* CCMP 331 (seawater) can be used as **biosensor** for the Biological Oxygen Demand (BOD), for example, the most pristine rivers will have a 5-day carbonaceous $[BOD]_{river} \leq 1$ mg/L, while, $[BOD]_{urban\ wastewater} = 20$ mg/L. Oxygen (O_2) is the terminal electron acceptor while hydrogen peroxidase (H_2O_2) an intermediate. **Oxygen Reduction Reaction** (ORR) in a MFC usually occur between 3 phases air (gas), electrolyte (liquid), electrode (solid) forming water or hydrogen peroxidase:

$$4\ H^+ + O_2 + 4\ e^- \rightarrow H_2O \text{ or } O_2 + 2\ H^+ + 2\ e^- \rightarrow H_2O_2$$

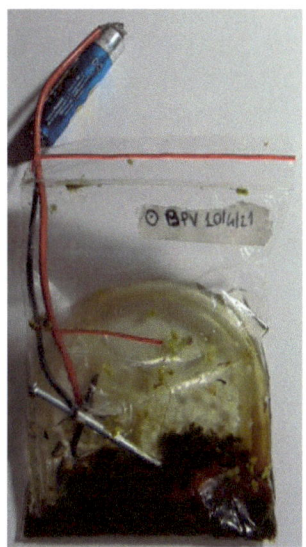

Figure 42 *Cyanobacteria* PhotoVoltaic zip lock pouches bags ($V_1 = 100$ mL, $V_2 = 30$ mL) with AA rechargable battery ($\Delta V = 1.5$ V), PolyEthylene (PE) salt bridge (ø = 0.5 cm), graphite rod (sp^2, $\Delta V_C = -0.4$ V) cathode and aluminium nail ([Ne] $3s^2\ 3p^1$, $\Delta V_{Al} = -2.3$ V) anode electrodes - Image Source: © Copyright Antonio Silvestro, 2021.

2D printed BPV cells with standard empty and disinfected (ethanol C_2H_5OH) cartridges onto A4 paper sterilized via microwaved in transparent plastic sheets for couple of seconds, transferred onto a red agar alginate substrate plate for incubation at T = 30 °C at an irradiance I = 20–30 µE/m²/s of fluorescent white light for t = 3 days.

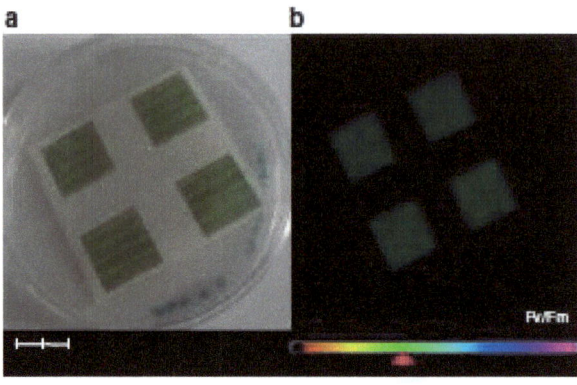

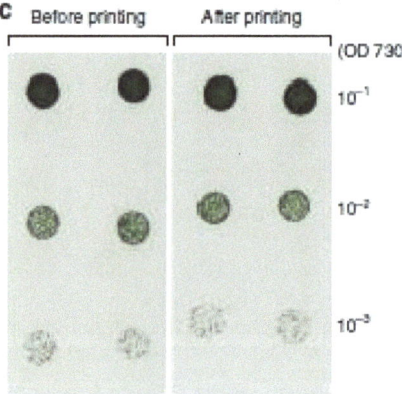

Fig. 1 Cell viability and photosynthetic capabilities of digitally printed cyanobacteria. **a** Photograph of inkjet-printed *Synechocystis* cells after 3 days of incubation. Scale bar measures 2 cm. **b** Chlorophyll fluorescence image of the sample **a** by imaging PAM, showing maximum quantum efficiency of PSII (Fv/Fm) at the values of about 0.4 according to colour gradient in the legend bar. **c** The panel compares the growth of *Synechocystis* colonies before and after the inkjet printing process, following 5 days of incubation on a BG-11 agar plate. A 3 µl aliquot of cells from a dilution series representing 10^{-1}, 10^{-2} and 10^{-3} of the original suspension was spotted. For the most dilute cell suspension taken after printing, 90.5 ± 10.6 colonies were counted, whereas 87.5 ± 12.0 colonies were counted before printing. The difference between these values was found to be not statistically significant (one-way ANOVA: $p = 0.815$) (Supplementary Table 1)

Figure 43 *Cyanobacteria* 2D printed photovoltaic cells - Image source: DOI: 10.1038/s41467-017-01084-4

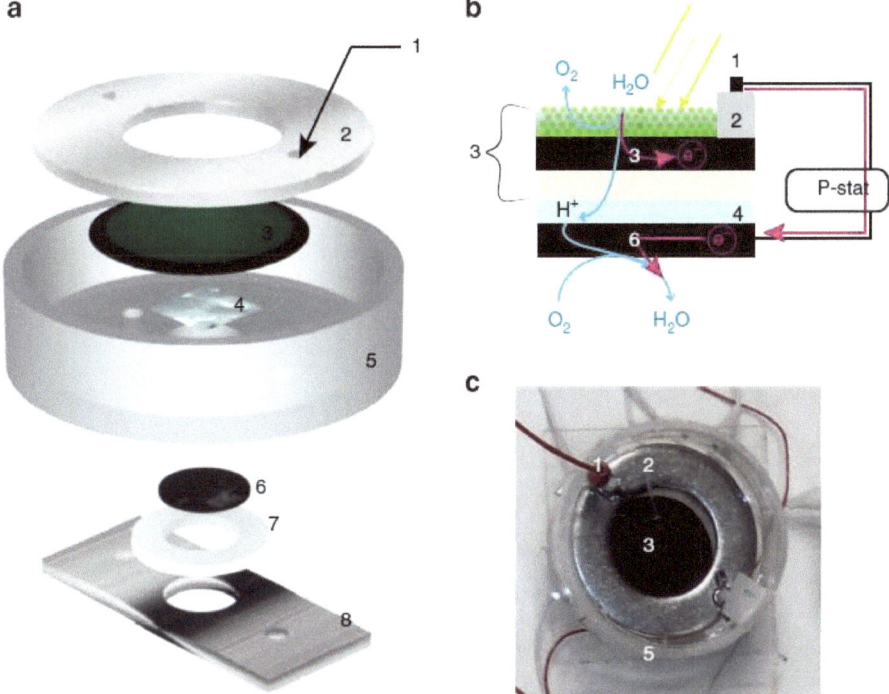

Figure 44 Electrochemical characterisation of a digitally printed bioanode in a hybrid BPV system 8T = 3.5 mA/m^2. Schematic representation (semi-exploded view) of the BPV unit with printed paper-based anode. Clamping screws (1); marine grade Stainless Steel (SS) ring for contacting the Carbon Nano Tubes (CNT) anode (2); printed CNT anode in black (ø = 6 cm) with printed photosynthetic microorganisms in green (ø = 40 mm) with a total area A ~ 28.4 cm^2 (3); hydrogel (4); Plexiglas vessel (5); carbon paper-Pt, with a total area A ~ 3.5 cm^2 was used as cathode (6); silicon (Si) O-ring (7), SS plate used to clamp all the components together (8). V ~ 60 mL of BG-11 medium 3.6 % (w/v) sodium chloride (NaCl), supplemented with 1.5 % (w/w) agar, placed above the printed cells in the chamber formed by the top plate – Image source: DOI: 10.1038/s41467-017-01084-4

Heterotrophic inoculum for Black Hole cathode (-)

Diverse viruses and bacteria could be tested for being teleported from a laboratory to another Fraunhofer Far-Field (FF) via biotech Wireless Power Transfer (WPT), perhaps, quiescent bacteriophages in lysogenic cycle parasitic cryogenic methanogens like ***Thermotoga*** *neapolitana,* or sporigens like ***Clostridium*** *botulinum* would succeed in the transposition. Bio-engineered ***Escherichia*** *coli*, human gut symbiont, could be used for teleportation trials before stepping to superior sentient beings.

Extremophile microorganism of the *Archaea* linage, precisely, psychotropic, anaerobic, cryogenic *Firmicutes* ***Leuconostoc gelidum*** *subsp*. 275 *gasicomitatum* and *L. gelidum subsp. gelidum*, dominant spoilers of Finnish, Belgian and Japanese *Ready-To-Eat* (RTE) meals, as elucidated viable count with Hight Throughput Sequencing (HTS) 16S rRNA and bioinformatics pipelines, could be used for the teleportation assay.

Alphaproteobacterium Magnetospirillum magnetotacticum **MagnetoTactic Bacteria** (MTB) secrete two main types of compounds in response to the displacement through the fringe between a medium lacking of oxygenated and one full of it oxygenated as during the *Great Oxygenation Event* (GOE) 2.4 BYA: magnetite (Fe_3O_4) and/or greigite (FeS_4), for which being LUCA anaerobe and H_2-dependent Fe-S ferredoxin reducer, it could have been in its genome preferentially the sequences for metabolizing the first. Precisely, LUCA should have been used Nitric Oxide (NO), nitrate (NO^{3-}) or sulfide (S^{2-}).

Furthermore, being the human heart cardiac muscle relaxation and even the fast vibrations of its syncytium tachycardia induced by NO, the descendant of LUCA could be present in the unoxygenated veins coming from the gut microbiome. Nevertheless, being the momentum of this ferromagnetic exudates related by the following equation: $p_{Fe3O4} \approx 3\ p_{FeS4}$, the external magnetic field due to CWB generator would let be magnetite properly for being utilized into the levitator rotor of *'Zeus - Genset (engine and generator)'* (Kindle eBook 1.48 € or Paperback 4.70 €) (https://www.amazon.com/Zeus-Genset-generator-Antonio-Silvestro-ebook/dp/B08D3WJ1PF/ref=sr_1_13?dchild=1&qid=1596535795&refinements=p_27%3AAntonio+Silvestro&s=digital-text&sr=1-13&text=Antonio+Silvestro) in its biotechnological version LUCA-based.

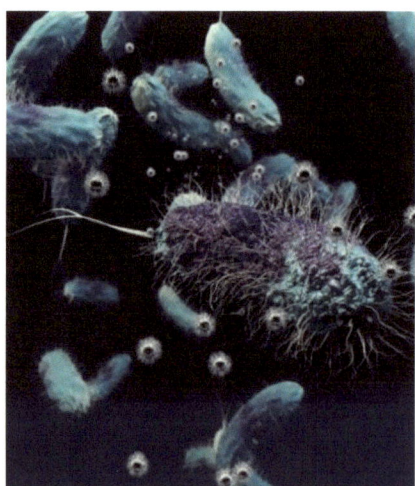

Figure 45 *Shewanella oneidensis* (ATCC® 700550D™ 311 € https://www.lgcstandards-atcc.org/en/Form_From_Scratch/Shopping_Cart.aspx) – Image source. Venkateswaran et al. https://www.eurekalert.org/multimedia/pub/108329.php

Electrogenic bacteria would be suitable to be used as models for the teleportation, being themselves used to be passed through by high electricity and potential and capable to transfer electron with cellular different structure that would prevent any damage to their nucleus and contained genome during the wireless quantum data transfer, destruction via the nirvana, construction via the *Sthula Sharir*, protecting layers guiding evolution.

Conductive **nanowires** like the conjugation T-pilus ($10 < \varnothing < 150$ nm, $L \geqslant 10$ μm) would facilitate the horizontal *Electron Transfer* (ET) to the anodic Black Hole (-) of the Einstein-Rosen bridge, have been found in the *Dissimilatory Metal-Reducing Bacteria* (DMRB) *γ-Proteobacteria Alteromonadales* **Shewanella oneidensis** MR-1 gram-negative heterotrophic facultative anaerobes (N_2 bubbling) in dark condition, thermophilic fermenting bacterium **Pelotomaculum thermopropionicum**, cyanobacterium oxygenic phototrophic **Synechocystis** PCC6803 in response to terminal electron acceptor limitation (O_2), (NO^-, SO^{2-}) and (CO_2), in continuous flow chemostat, sealed batch and solar bioreactor, respectively, correlated with the expression of the genes coding for the deca-heme *cytochromes* MtrC, OmcA, mshA, pilA and GspD, involved in the type II secretion pathway, and solid-phase Fe (III), Mn (IV) *cofactors reduction*, coordinated in poorly water soluble organic complexes with enzymes characterized by substrates such as pyruvate, lactate, acetate, fumarate and nitrilotriacetic acid current generating. For example, lactate ($C_3H_6O_3$) would undergo dehydration into pyruvate ($C_3H_4O_3$), itself decarboxylated into acetate ($C_2H_4O_2$) that hydrolyzed reacting with $2H_2O$ would release $2CO_2 + 8\ e^- + 8\ H^+$, later fixated by autotrophs (micro)organism in the environment.

Research Derek Lovley University of Massachusetts about *Geobacter* protein nanowire, *Shewanella putrefacians* and *Geobacter metallireducens* reduce Fe (III) to Fe (II), can be used to make electricity flow through them. *Shewanella oinendensis* in anaerobiosis can utilize *TriMethylAmine N-Oxide* (TMAO), *DiMethylSulfOxide* (DMSO) and oxidized metals such as Fe (III), Mn (IV) and U(VI), either as soluble complexes or within solid mineral (hydr)oxes, as terminal electron acceptor anode (+) (Marian Breuer et. al., 2018). The nano-polymers ca be visualized by topography obtained with Tunneling Spectroscopy (TS), redox tone assay - Transmission Electron (TEM), Scanning Electron (SEM), *Atomic Force* (ATM) and Scanning Tunneling Microscopy (STM) (PNASY A. Gorby et al., 2009). *Shewanella oinedensis* MR-1 (ATC 700550) can reduce organic compounds, metal ions and radionuclides under UV irradiation in aerobically growth Davis medium, aerobic and suboxic growth conditions in M1medium. *Shewanella spp.* could process, brake down the radionuclide *Uranium* U-235 and U238 combined with oxygen into Uranium Dioxide (U_2O) in the dye, but as they are not used together it wouldn't happen, on contrary the enzymatic complex of these bacteria would permit the metabolism of nuclear isotopes in the fluorescent Singularity. *Shewanella spp.* descendant of LUCA may have opened the nanopores of the Singularity escaping it breaking down the U_2O enzymatically, escaping that bright womb for diffusing within the protenaiceous medium. Both photoelectric and photovoltaic effect would be generated by the UV light diffusing from the uranium spontaneous fluorescent emissions of the Singularity as conical spiral super-current and condensed potential on the parabolic electrodes of the Black and White Hole.

Multi-haem, iron coordinated by protoporphyrin IX covalently linked to the peptide by thioether bonds (C-X-X-C-H motif), c-cytochromes ET involving respiratory enzymes through Inner (IM) and Outer Membranes (OM) up to tens of nanometers catalyzing the extracellular reduction of solid substrates, including electrodes and insoluble mineral oxides contributing to biogeochemical N, S, Fe cycles.

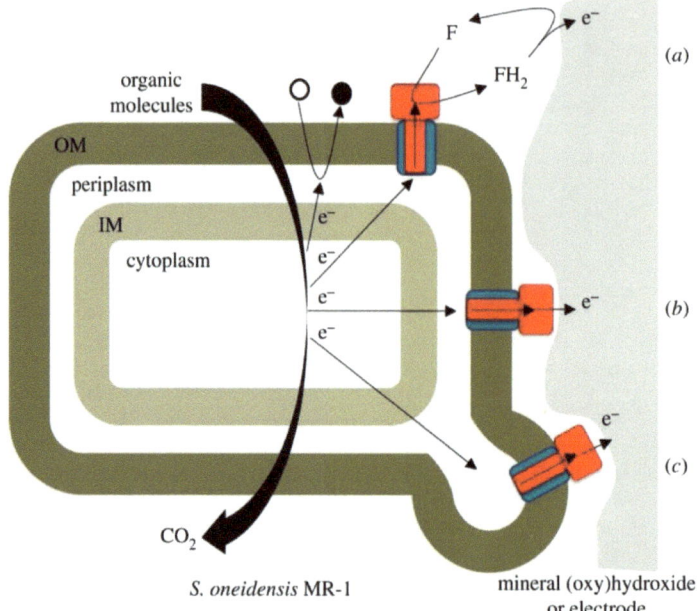

Figure 46 *Shewanella oneidensis* MR-1 growth in anaerobiosis oxidising organic matter at the anode (-) for elucidating the respiratory ET mediated by flavin via cytochromes and nanowires to the anode (+) – Image source: Marian Breuer et. al., 2018.

Heam-to-heam electron transfer rate is well described by **Marcus theory** in the non-adiabatic limit:

$$k_{ET} = 2\pi/\hbar < |H_{ab}|^2 > (4\pi \lambda K_b T)^{-1/2} \exp(-\Delta G + \lambda)^2 / 4\lambda k_b T)$$

Where:
ΔG = driving force, Gibbs free energy
λ = the reorganization of free energy
H_{ab} = electronic coupling matrix element

Shewanella oinedensis strain MR-1 **genome** (https://www.ncbi.nlm.nih.gov/nuccore/AE014299) have been sequenced by the Department of Genomic U.S. researcher functional noting 4.931 predicted Open Reading Frames, 40 % of which just hypothetical, profiling genes involved mainly in energy, ion transport, secondary metabolism and signal transduction.

Pilin building block of nanowire α + β protein characterized by a very long N-terminal α-helix, 3 β-sheet and 4-loops. (http://www.rcsb.org/structure/4D40) with electrical conductivity and adhesion properties, both within the microorganisms and between them and the electrodes proteins. Conjugation pili and fimbriae assembly mechanism monomers is not known, although chaperones have been identified (C.H. Jones et al., 1993). *Schumann radiation* ($v_0 = 7.83$ Hz) created by

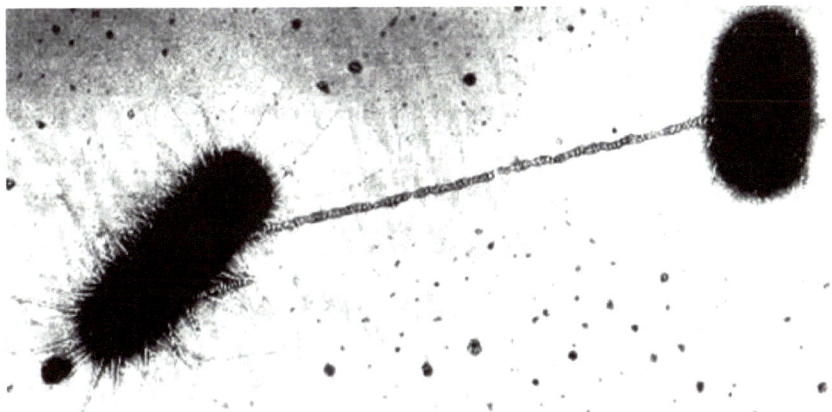

Figure 47 Conjugation pilus between *Shewanella spp.*

CWB generator amplifying the natural one in the empirical P-transition, induce the conformational change of the input domain of the sensor kinase in the plasma membrane of the gram negative *Shewannella oinedensis* activating the phosphorylation by the transmitter domain of the receiver domain of the response regulator that via another structural change inducing an intracellular signal cascade via MAPK till the nucleus where transcription activator factor favoring the expression of the PilA, E, X, V, GspG, H, I, J, MshA, B, C, D, O and FimT, U genes coding for the pilin nanowires and the excretion complex that leave them out of the cells.

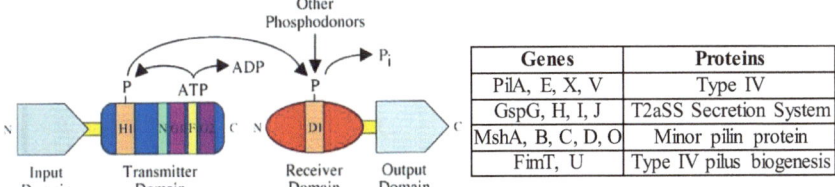

Genes	Proteins
PilA, E, X, V	Type IV
GspG, H, I, J	T2aSS Secretion System
MshA, B, C, D, O	Minor pilin protein
FimT, U	Type IV pilus biogenesis

Figure 48 Organization of a proto-typical two-component wireless regulatory proteinaceous system in whish is to note the conserved Histidine (H) at N-terminal of trasmitter and the Aspartate (D) at the receiver domain – Image Source: Jeril L. Appleby 1996, Parkinson 1993.

'Wormhole *in vitro*' © Copyright Antonio Silvestro, 2020

Being the preferential **marker element** for following the photoelectrical phenomena of the spooky particle at a Near Field (NF) as Fraunhofer Far Field (FF) distance in entanglement affected by quantum leap in the atomic orbitals would be the phosphate bond to the Adenosine (ADP, ATP, DNA, RNA) which wave function would carry the power wirelessly transmitted in the space always full of our ignorance. A transformer like a Solid-State Tesla Coil (SSTC) would accomplish the Wireless Power Transfer (WPT) that let fill the valence electron holes of the **phosphorus** $(_{30.97}{}^{15}P)$ atoms, lambents to the internal walls of the bulb Singularity, by the hot plasma current in the Ultra High Vacuum (UHV), filled with colling Helium superfluid, confined by the powerful magnetic field of Zeus, as in a Tokomak, by the superconductive Hera coil. Where in the photon transduction the info is duplicate from an Eppendorf to another, the Zeus Singularity with the Einstein twins within, otherwise, Ares magnet, time-travelling back on time against gradient of pressure and potential remain only one bouncing from the Black Hole to White Hole, or merely teleported in the future jumping on the opposite way. Imagine now the accretion disk of the two astronomical holes as the valence orbitals of an atom, well, you may understand that the nucleus is the masses body falling bending the Minkowski space time. Actually, genomic and environmental evolution see their intimate correlation due to this element that would let the nucleic acid variably express or silence themselves in relation to light in the free space filled with your wise measurements.

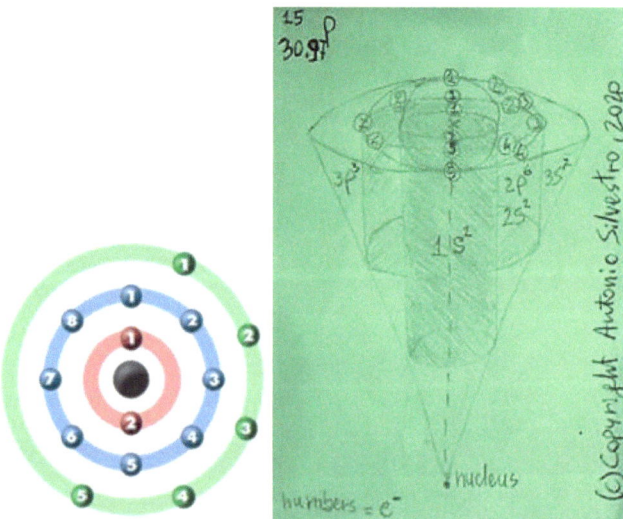

Figure 49 Phosphorus $(_{30.97}{}^{15}P)$ 'Niels Bohr model' (left), Black Hole-based 'Silvestro model' for which the electrons in the orbital $1s^2$ and $2s^2$ are in a common central spheroid within the a unique prolate ellipsoid $2p^6$ and $2p^3$, and the nucleus fall down to the vertex of an imaginary Dirac cone in which two inscribed cylinders would represent the volume in which the electrons would flow when escaping the most stable orbitals attracted by the nucleus, the closest deepest, nevertheless, never reaching it – Image source: © Copyright Antonio Silvestro, 2020.

As you may understand for the previously shown P-orbital there would be its image reflexed in the H_2O mirror, the one that could bring the atom back in time to far past in the White Hole, leaving just its nucleus in the shared present hypersurface.

One of the highest number of **sensors proteins** in prokaryotes have been found in *Shewanella oinendensis* (e.g., 45 histidine kinases, 26 methyl-carrier chemotaxis proteins and 52 diguanylate cyclase) make this a model organism for signaling and ultimately for a coordinating assembly.

Teleportation milestones

The fiber optic evolution began since the **photo-phone** (U.S. Patent 235, 199) invented by Alexander Grahm Bell in 1880 a device, similar to the modern telephone modulated by electrons into the circuit, for the transmission of radiation emitted by the vocal cords. Regenerable information can be transmitted via light pulses through optical fibers, with minimal attenuation and dispersion to reach immunity from interferences, for long distances and to pointed target destinations via RF U band ($1.63 < \lambda < 1.66$ μm).

Fiber internet is not distance sensitive like copper-based services – broadband, speeds up to $v = 1$ Gbit/s, over 10^{25} bit/Km s in the Bell laboratories delivers much faster files, as the Mathlab or Scilab electromagnetic grams and downloads than broadband.

By 2002, an intercontinental network of $25 \cdot 10^4$ km of submarine communication cable, which cross section from the cortex till the core is done of PolyEthylene (PE), Mylar Tape, steel, aluminum, water barrier, *PolyCarbonate* (PC), copper or aluminum, petroleum jelly, optical fibers, with an overall capacity of $C = 2.56$ Tb/s.

The information about the teleported should be sent in safety utilizing a kind of proxy using quantum computer networks done of a web proxy server acting as anonymous intermediary between two clients for maintain privacy. Quantum information, quantum state or qubit can be teleported, term coined by Charles Bennet, between a sender and a receiver in entangled – **Bell state**. Quantum bits can be transferred via entanglement as the pair of twins' corpuscles which state could be teleported, in eigenstate or superposition as electrons in the orbitals or thought between telepathic humans. In 1993 the Charles Bennett IBM research group demonstrate the **teleportation of the information** within a particle. In 1997 UV photons were teleported at the Innsbruck University, later on in 2004 Viennese physics teleported this **packet of energy** beneath the Danubio river via a fiber optic cable for l = 600 m. Three *beryllium* (^4_9Be) have been entangled by the researcher of the National Institute of Standards and Technology in Washington D.C., while in 2006 light beam carrier of information with *caesium* ($^{55}_{133}Cs$) were teleported for half a yard by Eugene Polzik team for the cutting-edge research project conducted by Niels Borh and Max Planck Institutes, furthermore, quantum information have been teleported from hyper-cold bosons (T = 3 K) with the lowest energy state *rubidium* ($^{37}_{85}Rb$) in a *Bose-Einstein Condensate* (BEC) state, pulse a light beam through an optic fiber beating with other BEC returning to the original state of matter over a distance of l = 150 meters using entangled photons by Xiao-Hui Bao at the University of Science and Technology of China in Hefei in 2012. Rb BEC could be immortalized onto mirroring parabolic sheets delimiting the accretion disks of Black and White Holes, let the light iso-beam pulse along the main axis of the Einstein-Rosen bridge teleporting the information content in the Juno globular superconductive $YBaCuO_2$ coil in the Jupiter Singularity coding in its valence electrons flowing in the electro-catalyst conical spiral etched into the Nafion Proton Exchange Membrane (PEM) of the Einstein-Dirac Cones which walls would be characterized by hyperbolic time delayed *luminescence* (T (t) ≈ 2 t/ t_{02} + t_{n2}) and *Poisson counting* statistic (P[n] ≈ [$n^k k_e^{-nk}$)/n!]. The Einstein-Dirac cone would spin in opposite directions, clockwise Black Hole (-), white, counter-clockwise White Hole (+) Kerr-Newman.

Included the *cosmological light horizon*, the maximum distance from which particles could have travelled to the observer in the age of the Universe, has no matter for quantum teleportation. The latest reported record distance for quantum teleportation in space is 1400 km by the group of Jian-Wei Pan using the *'Micius satellite'*. In 2008, M. Hotta proposed that it may be possible to teleport energy by exploiting quantum energy fluctuations of an entangled vacuum state of a quantum field even for *interstellar distance* more than light through vacuum ly light-year 9.5 · 10^{15} m energy transfer, the amount of teleported energy is non-zero, but negligibly small in ordinary silicon-based computer working in categorical pulsing bits (0 or 1), while, the teleportation protocol would be effective in using quantum computers of the same size that works in continuous sinusoidal 0 < qubits < 1.

Any border in the Bubble Universe is in causal contact as the skin of two confused, stoned lovers under a flood of Barolo wine of memory lost in the present hypersurface of their own heart, awakening the next day by the dopaminergic cloud of transcendental 25 mg/mL *Coffea arabica*. Cause and effect are on the membrane dividing the past and the future. The chemical information to process would move in an electronic device, through summatory series and multiplying parallel **quantum logic blocks** depending on the obstacle, the transitional gate to overcome (*'Quantum Computers Explained – Limits of Human Technology'* by Kurzgesagt – In a Nutshell Youtube channel), the grey hypersurface of the present between Black and White Hole, do not able to overcome with a Maxwellian view of the XVIII century, but with relatively more update integration of the general spacetime stretching the gate, making the leap of the Singularity possible via the folding of the two half of the Einstein-Rosen bridge in one only 3D cone or even 2D isosceles triangle more suitable for pragmatic engineering applications like the construction of solar batteries that instantaneously displaces electron-holes pairs along the fringe.

The primordial molecule bringing with it the memory of the origin of life, the crystalline **H_2O** of Masaru Emoto, changing in various conformation as the silica quartz to which the heart in geologically approximated would be the simplest molecule for evaluating the probability of its instantaneously heterogeneous respect the medium, isolated in evanishing sonoluminescent bubbles where is hydrolyzed or photolyzed, ionized and fissed leading to gamma rays from the molecule bringing memory to teleportable humans, done of about 75 % of it, in the Ultra High Vacuum (UHV) of their *Nirvana Sharir*. Extremely Low Frequencies (4 < v_{ELF} < 8 Hz, λ = 100 Mm, E = 12.4 feV) have effect on heat fusion period, hydrogen peroxide (H_2O_2) and oxygen (O_2) content of water.

The experimentation for transducing energy-matter from a place to another over Fraunhofer distances, perhaps, became when the immunologist Jacques Benveniste displaced **Ovalbumine** (v_{Ova} = 22 kHz ≈ $v_{lightenings}$, t = 1 s, 16 bits), the most abundant protein in plasma that bind water avoiding its diffusion into interstitial fluid (Mendelian inheritance Ovo-albumin OMIM) digitally amplified using LM Op-Amp and transferred through transatlantic, from Clamart to Chicago, the antigen (Ova), then perfused into heart isolates Ova-immunized *Cavia porcellus* (Benveniste et al. 1997).

F. A. Popp investigated to the **DNA non-linear photon trapping** as a coherent broadband PAR biocommunication system for which the beads on a string, Chromatin, resonate within the nuclear membrane in eukaryotes and inside the plasmalemma in prokaryotes as a quantum resonator, an exciplex emitting biophotons. Epigenetic unveil the existence of another code the histone one and so that the double helix designed by the intuition of J. D. Watson & F. H. C. Crick, R. Franlin & M. Wilkins is not alone, but in high dynamic chromatin state, that remodeling itself regulate gene expression. Walter Gilbert hypothesis where is neither DNA, but RNA the basal information of life for which entities, not properly organisms, could naturally be hosted by living photosynthetic organism like LUCA interfering auto-catalytically satellite with its growth and development.

DNA photon-transducer ideated by Luc Montagnier team (*'DNA waves and water'*, 2011, *'Transduction of DNA information through water and electromagnetic waves'*, 2014 and *'Method for digital Transduction of DNA in living cells'*, 2016 - Luc Montagnier et. al.). Audio Frequencies emitted (500 < v_{AF} < 3000 Hz) were detectable by high diluted water filtrates [(10^{-15} < serial dilution < 10^{-2} through filter of ⌀ = 100 nm, 10^{-13} < bacteria < 10^{-7} m (e.g. Mollicutes *Mycoplasma*

'Wormhole *in vitro*' © Copyright Antonio Silvestro, 2020

pirum) and 10^{-10} < viruses < 10^{-6} m via net of ø = 20 nm (e.g. FeV, Hepatitis, influenzae)]; among this 2 ng/mL samples of short infectious Human Immunodeficiency Viruses (HIV) {Lentivirus ssRNA-RT group VI, capsid (ø = 120 nm), overtime cause Acquired Immuno-Deficiency Syndrome (AIDS), 9 < age survival estimated < 11 years, sexual or parenteral breast milk transmitted, HIV-1,2 infects $CD4^+$ T_H via micro-tubular transport integrating its two epistatic genomes (composed of 1000 < NTDs < 2000) into the target host cell's nucleus - lysogenic cycle characterized by a latent phase, killed by $CD8^+$ T_c, via (pyro)apoptosis, macrophages, microglial and dendritic cells} DNA human lymphocytes Long Terminal Repeat (LTR) = 104 bp marker sequence were chosen and placed in Eppendorf tubes, surrounded by an inductive and conductive copper solenoid (Z = 300 Ω) in a cylindrical tube internally lined with ferromagnetic mu-metal sheet (77 % Ni, w_{Ni} = 1 mm), linked to a generator of a clean background geomagnetic fundamental Schumann radiation of ν = 7.83 Hz, amplifying with a microphone and recording with a sound board in a laptop computer (e.g. Lenovo X240 CPU: Intel inside CORE i5 Microsoft Corporation Windows) with an oscilloscope software (e.g. Scope C. Zeitnitz 2005-2015, graphs in .csv and audio in .wav format), the first 6 seconds = 6_{th} qubit output signal after 16 < t < 18 h in another distant Eppendorf that were just filled with water V_{H2O} = 1 mL. The corroboration of the DNA photon transduction, otherwise, teleportation of the genomic information via the electromagnetic radiation has been done adding DNA Polymerase Chain Reaction (PCR) reagents to the signalized water and separating the thermocycler amplicons in an agarose gel electrophoresis detecting the expected LTR fragments (98 % identical). Rearranging the Luc Montagnier DNA photo-transduction, in which DNA-polymerase DNA-dependent (e.g. *Thermophilus acquaticus* Taq polymerase) assembles the acid bringing the information of life far field the substrate in water solution, with Stooling C. Lartigue genome transplantation would be possible transplant multiple genes in one only empty host, each characterize by a unique wave function in a complex genome teleportation guided by electromagnetic radiation in absence of gravity with strong and weak forces playing a role in the DNA scaffold and nitrogenous base pairs bonds, respectively.

As said by Daniel Winter, **DNA** (18.5 < $ø_{DNA}$ unfolded < 25.5 Å) emits an EM field that goes faster than light at tachyon speed, potentially travelling into silica optical fiber capillary (20 < $ø_{intern}$ < 100 μm, pH < 3 positively charged), which oscillatory **wave in the WormHole in vitro** cylinder would be theoretical representable by the **radius equation** as the sum of the wave functions of all the electrons:

$$\nabla^2 A + K^2 A = x^2 d^2y/d^2x + 2x\, dy/dx + [x^2 - n(n+1)]y = 0$$

Where:
A = radiation amplitude
K = wavenumber

Nucleic acids in the universal medium would have being teleported by light quanta through the oscillating CWB since the Omega Centauri in the centre of the Milky Way Galaxy formation.
Properly isolating couple of endo-symbiotic unicellular descendant of LUCA, free-living in mutual symbiosis or *quorum sensis* colonies could be among the plethora of example of spontaneous quantum transfer that could be reproduced *in vitro* for simulating **biological teleportation** of energy-matter suitable as model for superior organism like *Homo sapiens* and unofficial higher *Homo atm*. Certainly, a step forward in the biomedical trial and the capability to teleport not just unicellular, but also colonial and pluricellular organisms would be counting the Colony Forming Units (CFC) with algorithm proportional with the mass of the single cell. Perhaps, isolate STEM cells of two lover Far Field, that feel each other could be a way for demonstrating the electromagnetism, quantum mechanics and general relativity behind the mental body *Manas Sharir*.

Wave phylogenetic three from mother to fetus, then adult and finally consort, using the quantum of light entangled under the Schrödinger electron equation could redesigned.

Quantum mechanics could be utilized in **Next Generation Sequencing (NGS)**, for example, instantly amplificated the target nucleotide fragment between two coincident entangled Illumina pair-reads placed in two distant bioreactor cells. To convert the DNA photon transduced from an amplification process to a real teleportation, the phenomena should be reviewed as a resonating treadmilling process characterized by the balance between polymerizing/depolymerization, pushing the equilibrium for regenerating the donor DNA in the recipient membrane and degenerating the emitting information.

Till now have been successfully teleported photons, atoms and some little organic molecules. But why do we stop here? If life is nothing, but code that can be packaged, emailed, downloaded, and copied, why not use the same technology to transmit life? (Craig Venter, inventor of the *'Synthetic Biology'*). A recent experiment may have placed living organisms in a state of quantum entanglement (Jonathan O'Callaghan on Scientific American October 29, 2018). As you may understood, quantum superposition of small organic molecules, their electrons and the signal processable by *Integrated Circuits* (IC) can easily be transduced as data along the silica optical fibre of the Wi-Fi, but what would make the change toward a biological teleportation is the utilization of the synthetic biology and cloning, using empty cells of ***Mycoplasma** genitalium* or *M. pirum* as host of the waveform into their loop membrane. **Ehrenberg–Siday–Aharonov–Bohm effect**, is a quantum mechanical phenomenon in which an electrical charged particle affected by an electromagnetic potential, despite being confined to a region in which both the magnetic field (B) and electric field (E) are zero are free to move, is due to coupling of antenna complex of the photosynthetic pigments of LUCA representative descendant which amino acids have been expressed in the water in an Eppendorf far field instantaneously via Faraday induction and electron stripping due to the oxidative action of the magnetized H_2O. The multi-signals containing all the information about the microorganisms should travel through many chamber pair as light does in the antenna complex of the photosystem for which cyclic excitation quantum emission would be through the processing of the data. Hence, not just magnetic field as the Spanish trio transferred through the Einstein-Rosen bridge, and being possible to photon-transduce DNA as Luc Montagnier demonstrated, the teleportation of a microorganism in a WormHole *in vitro* form the negative to the positive it would be realizable using an empty hosting Singularity, the plasmalemma of a microbe like the *Mycoplasma* used by Greig Venter or anucleate Red Blood Cells (RBC).

Figure 50 Iso-thermal *in vitro* recombination of the first bacteria have been synthesized with a BioXp 1st DNA printer (DBC Prototype) - Image source :
https://www.ted.com/talks/dan_gibson_how_to_build_synthetic_dna_and_send_it_across_the_internet/details#t-310707

Digital-to-Biological Converter (DBC) for fully automated, versatile and demand-based production of DNA-RNA and protein sequences for virology and immunology applications (Dan Gibson). Brain and Quantum Computers could be set in an entanglement state for transferring digital information into the living organ. Trial after trial, from the most inferior organism to the superior *Homo sapi...*'ops' *Homo atm*, send DNA in the lymphatic system of human embryos initially private of their genome using cytologically transplantation like the once used in the Fertilization in Vitro and Embryo Transfer (FIVET). Developing a multi-factorial matrix made of screened values obtained from the sensorial **biometric panel** analysis using the combinatory calculi to better define the digital information of the target to teleport, would permit its displacement to other place, not just itself, but even a better copy of the original teleported.

A new paper from a group at the University of Oxford is now raising some eyebrows for its claims of the successful **entanglement of bacteria with photons** led by the quantum physicist Chiara Marletto and published in October in the Journal of Physics Communications, the study is an analysis of an experiment conducted in 2016 by David Coles from the University of Sheffield and his colleagues. In that experiment Coles and company sequestered several hundred anaerobic photolithoautotrophic (photolyzed electron donors H_2, H_2S, S, Fe^{2+}, bacteriochlorophylls RC P808, 840, 795, 720, 750 send the electrons to the cyt 551, followed by reverse Krebs cycle for CO_2 fixation) green sulfur bacteria *(Chlorobi Chlorobea Chlorobiales Chlorobiaceae)* between two mirrors, progressively shrinking the gap between the mirrors down to a few hundred nanometers (l_{NF} = 100 nm). By bouncing white light between the mirrors, the researchers hoped to cause the photosynthetic molecules within the bacteria to couple or interact with the cavity, essentially meaning the bacteria would continuously absorb, emit and reabsorb the bouncing photons. It appears certain photons were simultaneously hitting and missing photosynthetic molecules within the scattering bacteria within the medium. Green sulfur bacteria reside in the deep ocean where the scarcity of life-giving light might even spur quantum-mechanical evolutionary adaptations to boost photosynthesis. Purple Sulfur Bacteria (PSB - *Gamma Ptroteobacteria Chromatiales*) should act similarly to the green in entangling quanta, for which the trial will be continued by the team of researchers led by Simon Gröblacher of Delft University of Technology involving more complex superior organisms: **tardigrades**.

'Wormhole *in vitro*' © Copyright Antonio Silvestro, 2020

Microbial Teleporter (MT) design and architecture

Electroactive microbial electrolysis, desalination and electro-synthesis cells are some of the techniques to use waste biomass usage to promote competition and transparency to reduce pollutant emission and improve the worldwide market fluidity, natural gas and electricity supply. **Microbial Fuel Cells** (MFC) can be used as sensor indicating the toxicity level of chemical compounds in wastewaters, rivers and seafloors. In this last, a potential difference can be generated by bacteria between the sediment and the aqueous phase below, precisely, sulpho-bacteria oxidize the carbon in the lees producing sulphide, then oxidized by other bacteria along the ocean food chain to the anode (-).

Sewage digestate from animal farm effluents and urban wastewater treatment are built in conformity and proportionally with volume and depuration capacity as ruled by the Ministry for Environment, Land and Sea Protection of Italy (MATTM) for water pollution tutelage as deliberated in S.O.G.U. n° 48 21/02/77. Furthermore, municipality landfilling leakage must be reduced of 65 % till 2030 according to the "Closing the loop – EU plan about Circular Economy" (COM/2015/614).

Economic incentives for producing to put greener products on the market, support recovery and recycling schemes (e.g., packaging, batteries, electric and electronic equipment's, vehicles) are needed. In accordance with initiative on waste to energy in the framework of the Energy Union 2016, and following the Guidance and dissemination of best practice on the cascading use of biomass and support to innovation in this domain through Horizon 2020, ensuring coherence and synergies with the circular economy when examining the sustainability of bioenergy under the Energy Union (2016).

Responsible and efficiently boosting livelihoods, nutrition and welfare for a sustainable word where a hygiene food safety, but also rapid accident treatment is effectuated under Standing Committee on the Food Chain and Animal Health (SCOFCAH) values using a Modus Operandi proper to manage food containing chemicals and chemical packages. Plastics container made according the Directive 94/62/EC, not just for food, but also for sanities, CDs and biberon, from PolyEthylene Terephthalate (PET) CAS: 25038-59-9 to the PolyAmide (PA) CAS: 32131-17-2, calibrating laboratory instruments to synthesize, use and monitor according to international standards materials registered via Joint Research Centre (JRC) authorized brands distributors (BCR®, IRMM and ERM®), new substances: colors, preservatives, antioxidants, acidity regulators, thickeners, stabilizer, emulsifiers, pH regulators, anti-caking agents, flavors enhancer, antibiotics, glazing agents, gases and sweeteners, additives, extraction solvents or monomers used for polymer production.

The global microbial fuel cell market is expected to reach 15 million € by 2025, at a **Compound Annual Growth Rate** (CAGR), constant rate of return over the time period (= Exponential growth, t = 1 year), of 9.5 % from 2017 to 2025 [("Microbial Fuel Cell Market, By Industry (Agriculture, Healthcare, Food & Beverage, Government & Municipal, and Others), By Region - Global Forecast to 2025" – www.micromarketmonitor.com MI 1000)].

$$CAGR\ (t_0, t_n) = [(v(t_n)/v(t_0))^{(1/t_n - t_0)}]^{-1}$$

Where:
t_0 = beginning time [s]
t_n = umpteenth time [s]
v = value

Microbial Fuel Cell (MFC) are biological transducer of chemical energy into fuel using *Prokarya Bacteria* (e.g., Robigus *Fungi Saccharomyces cerevisiae* used by Bacchus, *Shewanella oneidensis*

ΔV = 35 V, I = 2 mA) and oxidising agents. Anaerobic bacteria in Microbial Electrolytic Cells (MECs) transfer the electrons (e⁻) from oxidized substrate to the anode (-), while the cathode (+) is exposed to the oxygen (O_2), forming H_2O (ΔV ≈ 0.5 V). **Bio Electrochemically Assisted Microbial (BEARM)** cell by adding small amount of voltage difference (ΔV = 0.25 V) produced by bacteria at the anode (-), avoiding oxygen (O_2) at the cathode (+), producing pure hydrogen gas $(H_2)_{(g)}$ at these electrodes. Theoretically is needed a potential of ΔV = 0.41 V to make H_2 from acetate, but *Acetobacters* spp. can generally produce it with just ΔV ~ 0.25 V. Hence, to generate hydrogen, an external voltage power supply at list of 0.2 < ΔV < 0.8 V, under standardize controlled conditions (pH = 7, T = 30 °C, p = 1 atm), for an average electrolysis output 1.2 < ΔV < 1.8 V. *Systemic Hydrogen Yield* $(Y_{sys}H_2)$ = 4 mol H_2 : 4 mol $C_6H_{12}O_6$: 8 mol $C_2H_4O_2$, using power supply P = 100, 240, 1000 W, $ΔV_{DC}$ = 12, 24, 48 V.

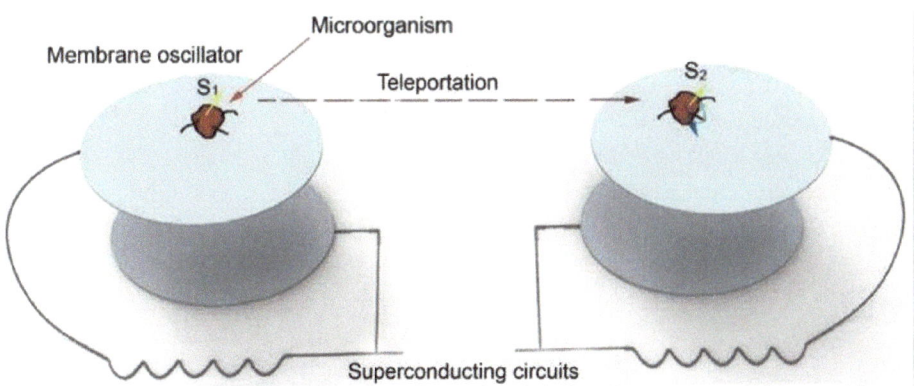

Figure 51 Quantum superposition, and state teleportation of glycine of a cryopreserved lighter weight microorganism on an heavier electromechanical aluminium ($\varnothing_{Al}$ = 15 μm) LC oscillator cooled at ground state into a superconductive circuit in entanglement inducing microwave field – Image Source: https://doi.org/10.1007/s11434-015-0990-x

Two microorganisms divided by binary scission placed on two oscillatory membranes representing the astronomical Holes would carry their quantum mnemonical state along vibrational waves superimposing in a steady state exchange their position in an instant where the knowledge of their qubits emerges simultaneously by both counterparts. In other words, placing microbes onto an undulating membrane of a subwoofer (A) would let them oscillate coherently with its radiation and levitating within magnetic lines and roto-traslating along the waveform generated, would reach another distant subwoofer (B) instantaneously. Quantum theory to exchange the whole genome between two really similar communities of micro-organisms in entangled (e.g., two strain of the same species, that differ just for a fluorescent marker gene) oscillating their electrons under two Coherent Domains (CD) at the unison. According to the **Anderson-Higgs-Kibble mechanisms** a photon entangled with the microbes is unable to leave the CD, because acquire an imaginary mass. The division spindle that may have been led an unicellular totipotent LUCA to differentiate in the tempest after the great collision between the Bubble Universe that we all know and the spot in the dark of your past, apparently unreachable, in your subconscious still to be awaken would come into you merely thanks to the folding, phenomena for that let the transposition them then to what there is now, the colorful nuances of energy-matter done of the rainbow remnant of that sparking impact that changed the destiny of living and not. Merely, the microbes could be placed on the same oscillatory aluminum membrane, but on opposite sides in the superconductive circuit generating CWB field enclosed in ferromagnetic electromagnetic shielding Mu-metal (Nickel-Iron : Ni-Fe) for being teleported from a cell to another.

Singularity's genomic matter photon-transduction through the empty Einstein-Dirac conical cavity paired which **Hamiltonian coupling strength,** characterized by the unitary ratio $g/\omega \approx 1$ (https://doi.org/10.1038/nphys3906), between the super-quantum resonating LC electromechanical oscillator Black-White Hole circuits coupled via Josephson Junction (JJ) as in super-conducting Quantum Interface Devices (SQUIDs), Magnetic Resonance Imaging (MRI) and MagnetoEncephaloGraphy (MEG), locking their two super-conductive waves of the Wormhole in phase, weakly linked in the vertex, where supercurrent (I_s) in cooper pair and voltage across (U(t)) are equals to:

$$I_s = I_c \sin(\varphi) \approx I_c$$
$$U(t) = \hbar/2e^- \, \delta\varphi/\delta t$$

Where:
I_c = critical current [A]
$0 \leq \varphi$ = phase difference of the super-conducting wave functions between two electrodes $< 2\pi$
$\varphi_{min} = 0$ Josephson energy minimum, ground state
$\varphi_{max} = \pi$ unstable and corresponds to the Josephson energy maximum
φ = quantum magnetic flux $= 2.07 \cdot 10^{-15}$ [Wb = T · m^2]
h-bar = reduced Max Planck constant $= 1.05 \cdot 10^{-34}$ J · s
e^- = electron charge = - 1

The Black Hole following the Supernova of the Solar System (SS) should have been symmetric with a White Hole from which descendant of LUCA found on Earth developed in a water mirror.

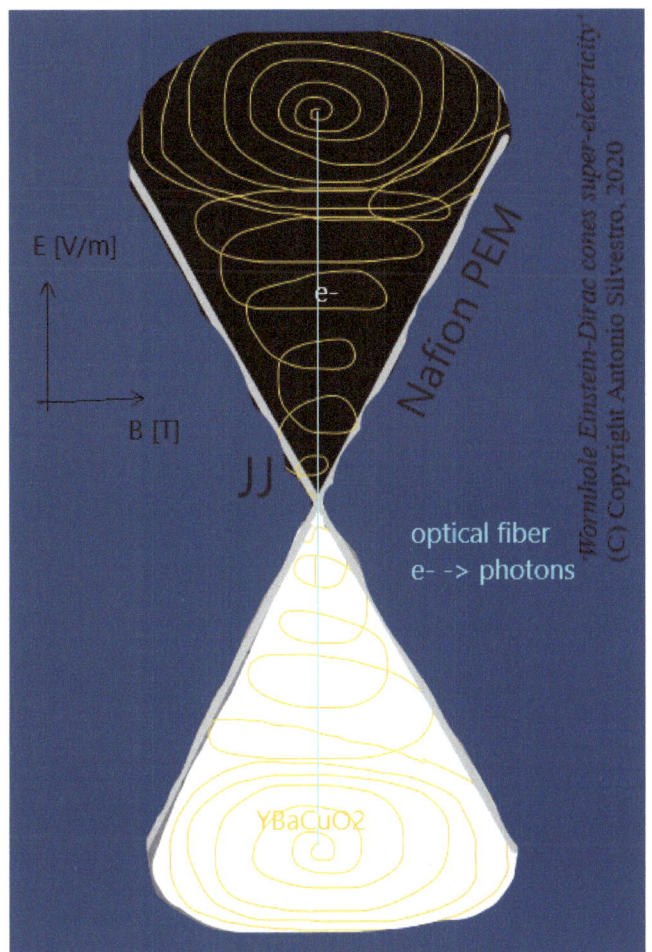

Figure 52 *'Wormhole Einstein-Dirac cones super-electricity'* done of biconical superconductive spiral (e.g., yttrium baryon copper dioxide YBaCuO$_2$) deposited into Nafion PEM cones affected by photoelectric effect and Archimedean spiral as floor plan, before the gravity fall onto the Josephson Junction – Image source: (C) Copyright Antonio Silvestro, 2020.

LUCA descendant in the **Microbial Teleporter** (MT) would be in Ultra High Vacuum (UHV) selectively inflated of inert noble gases (He, Ne, Ar, Kr, Xe, Rn or Og), hydrogen (H$_2$) and 4-helium superfluid (4-He), ferment oxidating the sugars in the Gaia geology relict medium, lyophilized remnant nano-powdery faces or the emptied human gut where LUCA descendant rhizobium thrive, let release lepton electrons (e$^-$) from C(H$_2$O)$_n$ metabolized into the Black Holes cathode (-) folded onto the White Hole cathode (+) from where they are naturally ejected or traveling in the superconductive cylindrical spiral compressed-relaxed by the Minkowski spacetime changing configuration from perfect cylinder to double Dirac-Einstein cones, out from the bottom cathode (-) into the Si$_2$O-aerographene Universe Bubble sphere where reach just two order less the speed of light, Bohr quantum leaping and Einstein photoelectrically emitting boson photons breaking the bubbles under the synthetic Cosmic Wave Background (CWB) shouted by the

'Wormhole in vitro' © Copyright Antonio Silvestro, 2020

Goddess *Juno* where giving birth to *Mars*, our new host planet, the God of war that who live in a peace welfare honor working and hard loving within his son *Cupid*. The CMW would have been spinning the dipolar Singularity at 21 billion times/s, let it oscillate from amplitude A_{min} to A_{max} and *vice versa*. The Singularity would be dielectrically heated by the Cosmic Wave Background at T_{CWB} = 2.75 K, exothermically rotating the dipole into the magnetic Einstein-Dirac conical cavity, smaller compared with its radiation wavelength (Extremely High Frequency EHF ν_S = 300 GHz > ν_{CWB} = 160 GHz). The upper limit of the Radio Frequencies (RF) concealing the radiation scattered from the insulating Si-Ti-or UO_2 aerographene shell of the Jupiter Singularity would see itself under fission reaction with Fe potentially magnetized, it would corroborate the photoelectric effect happening onto its peripheral surface from which electrons in motion could reach the speed of light, according the Hall effect, in a bluish monochromatic radiation, what, should precisely occur in the tangential points (T) between its circumference and the Einstein-Dirac cone sides. The background microwaves would have been access to the center of the Singularity, because in this case it would have been sterilize overheating release exothermically the minimum theoretical energy of about $E_{min} \approx 4$ kJ. Furthermore, it would inductively polarize with its permanent magnetic field the originary spheroid let it move counter- or clockwise in its rosette displacement. Water heated by the mm-waves shows *paramagnetism* as the orientation of the love stories of Zeus father of Ares, instead, associated with hydrogen-helium *decay* series and H^+-bonds, this last increasing under the raising magnetic field induced by the CWB in the Bubble Universe that would lead to slower passage through the cone's vertex. Teleportation may be accomplished let flow Fraunhofer Far Field electrons striped-out for the phosphate backbone of the DNA of LUCA descendant quantum leaping in the super-conductive $YBaCuO_2$ holes of the coil deposited into the Nafion Einstein-Dirac cones photo-electrically emitting quanta of light, bit after bit, in parallel networks managed by quantum computers ICs to which 'CWB generator' is connected for parallel Einstein-Rosen bridges teleportation where the intrigue network of telecommunication would fill all the Bubble Universe.

You here reading, or maybe loving one like him that would teleport your brothers and sisters over the distance of your mother Earth via the α-quartz silicon of your lovely piezoelectric heart newer affected significantly by the gravity that let fell the SuperNova (SN) that let us be. In the *Viññāṇa* there are always two ways, merely, attraction-repulsion according the classical physics, and the confined ionized plasma into your veins and arteries and in the inferior beings homophyletic to you back to LUCA , the magnetic field (B) generated by the wormhole that let Juno scream during the Big Bang perpendicular to the electric field (E) by her self-generated wile squeezing your singular heart would flux from low to high potential electroplates of the synthetic Einstein-Rosen bridge staging the originary birth. The compressive effect that let displace the Singularity probabilistic through the White or the Black Hole, should depend on **fission** exergonic reaction (D + T -> He + n_0 + γ rays) that let expand a thin layer of *PolyStyrene* (PS) foam (Styropyrene) between the super-conductive $YDaCuO_2$ and Ni mu-ferromagnetic bending progressively the inner cylinder into a Einstein-Dirac cone with its radial increment up to the limit that would correspond in size to the diameter of the originary spheroid ($r_{PS} = r_{Singulary} = ½ r_{Shwarzchild}$) stabilized when helium would become superfluid at $T_{4\text{-}He}$ = 2.7 K. **Synthetic Singularity's volume** would be equal to : V = 4/3 π ½ $r_{Schwarzschild}^3$ = 4/3 π ½ $(2 GM/c^2)^3$ = 48 m³, 3 < M = Black Hole Mass < 10 $M_\odot$, G = 6.674 · 10^{-11} m³ · kg^{-1} · s^{-2}, c = 3 · 10^8 m/s. David Layzer affirmed that the present moment always contains an element of genuine novelty and the future is never wholly predictable. Because biological processes also generate information and because consciousness enables us to experience those processes directly, the intuitive perception of the world as unfolding in time captures one of the most deep-seated properties of the Universe. For which as not be surprise full that in the next future may be possible to teleport information about the genome sequence of simple unicellular organism from a polarizing Einstein-Dirac cone to another joint to it via the apex in a foldable configuration bending within the Minkowski spacetime overlapping the accretion disks of the White, with Black Hole for an instantaneous quantum displacement of a harmonic function in an Euclidean space

'Wormhole *in vitro*' © Copyright Antonio Silvestro, 2020

using Cartesian spherical or cylindrical coordinates, which orientation changing can be represented with the Laplacian operator elegantly showing the probabilistic gravitational fields in the free-space of the Bubble Universe, otherwise, reduced to the heliosphere, none for a continuous unidirectional flux of charged particles like the electrons from the Black (-) and White Hole (+) for travelling in the past looking for the origin of the whole we know still now, on contrary teleport biological matter and its energetic content in the next instant. Universe has memory, if you leave an object in a place you should remember where is it, isn't? Well, is right for this that its information sealed in yourself quantum witnesser experimenting teleportation experiment, it would be able to teleport it imagine back to the reality left behind. Well, it may appear complex, but it isn't. Just think the microbes onto a capacitative oscillatory membrane leave under a continuous oscillatory field as the light itself according Albert Einstein thought to us. Well, within with it in the Black Hole, saturated of it, the microbes would absorb it all, spontaneously emit light in a continuous positive feedback within the empty medium for later jump in the past scattered everywhere, otherwise, its DNA would be canalized through fused non-crystalline silica-Nafion Einstein-Dirac optical fibrous cones teleporting its life information carried by the H_2O vapor of the medium. Dye-sensitized Erbium ($_{68}Er$) thermoplastic insulator [10 PolyVinyl Chloride (PVC): 1 SiO_2, TiO_2 or UO_2] in the Singularity walls could be used for marking the teleportation pathway of the spheroid, otherwise, could be highlighted a reference sequence of the descendant LUCA would be facilitating by the oxidation due to the presence of Fe-S proteins assimilating amino acids.

The holomorphic transition Atlas 6-dimensional Calabi–Yau manifold characterized by a number of coordinates specifying any point, open unit disk (D(P) = {Q: |P-Q| < 1} in C^n), between a Singularity to the contiguous of a parallel Universe via the Black Hole negative side (-) would generate violetish hydrogen (H_2) [75 % of the elemental mass of the Universe, which weight has been determined by Max Planck using the density of eigen-frequencies fundamental constants as: $m_H = 1/N = 1.62 \cdot 10^{-24}$ g, where: N = equivalent mass)] in higher amount with an acid catalysis lowering the pH using phosphate buffer $3 < PO^{3-} < 50$ mM, which powder bordering the perimeter of the Singularity could emit UltraViolet (UV), and $[NaH_2PO_4^2] = 4.5$ mM suitable for the backbone of the nucleic acids, $[NH_3Cl] = 28$ mM for the ammonoids chains, ferric nitriloacetic acid $[C_6H_6FeNO_6] = 0.1$ mM for the porphyrinic ring, sodium selenate [NaSe] = 0.001 mM for sensing the no-self, and acetate $[CH_3COOH] < 50$ mM for the fatty acids synthesis related to the Wood–Ljungdahl pathway found in LUCA and sodium lactate $18 < [NaC_3H_5O_3] < 30$ mM for stopping the lactic fermentation favoring H_2 Fe-S ferredoxin reduction leaving holes (H^+) in the medium surrounding the transparent silica α-quartz filled by the electrons flowing though the superconductive Juno coil, that compressed by the Minkowski spacetime would strip-out the e^- of their orbitals in a chaotic coherent colorful eigen-mode at the unison centered along the fundamental resonating frequency at ν = 1.3 GHz, vibrational string of the coherent atoms involved in intra/inter-molecular bonds, under polarizing light beams flipping rotating around an axis, changing the conformation of the *Anahata chakra* according the **M-theory** by Edward Witten (1995) elegantly unifying quantum mechanics with general relativity's gravitational force in a mathematically consistent way where different perspectives in a multidimensional Universe are highlighted "strings" seen in three-dimensional space as matter, light or gravity. But, when love is divided into good and bad by the father teaching the daughter or the son, the state become two one on another, superimpose. This is not an issue concealing infinitesimal small corpuscles, but also what they leave from their shah as the pastel coloring the white cellulosic paper leave traces, so the primordial elementary particles draw by moving along elliptical orbits the cylinder themselves, for keep moving in two superimpose Einstein-Dirac cone spirals, the White and the Black Hole depressed and relaxed by the curving Minkowski spacetime. No difference between seen and seeing in a relativistic view of the quantum complexity of cyclical astronomical phenomena that carry evolution of soul, soulless and non-genomic matter.

The information between at least two **chakra strings** in opposite anthropomorphic Black and White

Holes require the minimum number of substitutions called *Hamming distance* for let the complex excitation flow through the humans, waves through the CWB waving the Kundalini network as for example the three Moira does with Zeus in a divine familiar reconnection.

The ElectroMagnetic (EM) interference during the teleportation would be blocked by the meshes of a **Michael Faraday cages** (1836) rectifying the transition in the free medium, where hydrogen (H_2) meets helium (He). A hydrogen **transitioning** among two quantum energy levels LUMO -> HUMO determining the photon emission which energy is equal to:

$$E_{photon} = h\nu = E_2 - E_1$$

Where:
h = Max Plank constant = $6.6 \cdot 10^{-34}$ J s = $4 \cdot 10^{-15}$ eVs
ν = frequency [Hz]
E_1 = energy ground state [J]
E_2 = energy exited level [J]

Time-lapse video micro-camera and internal sensing probe connected to an Arduino UNO board for monitoring abiotic factor recording inside the MT cylindrical squeezable by the reduction of the outer pressure adding an UHV pump to the bioreactor filmed in its processing by an external video camera. The MT cylinder would have a cleavage furrow ring equator SiO_2 aerographene insulated contiguous with the Bubble Universe, for which the UHV variation in it and the spaçetime Minkowski compression along the hypersurface of the present inside the cylinder would let to the formation of two Einstein-Dirac cones, leaving the vertex among them insulated, while the cones superconductive along the spiral raising of the electrocatalyst (e.g. Cu, Ag, Ti, W or $YBaCuO_2$ used in the artificial WormHole of the Spanish trio) cyclotrons roto-traslating unfolding the telescopic sleeve would be confined by the coating super acid polymer film electrolyte ionomer **Nafion**® PEM (sulfonated TetraFluoroEthylene, 424 DuPont, CAS Number 66796-30-3, 1 W < P < 500 kW, 60 €/W), which thickness depends primarily on the differential pressure (Δp) expected across the membrane, [minimum Nafion 211 (2.5 cm), thicker for very high Δp applications (Δp > 1000 psi = 68 atm) or very long operating life (t = 10^5 h, 11 ½ years)].

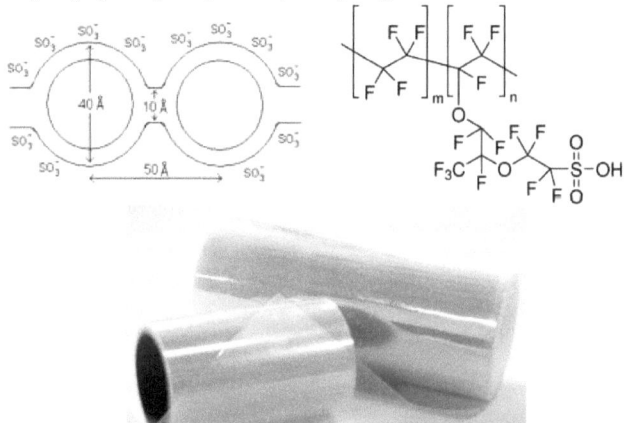

'Wormhole *in vitro*' © Copyright Antonio Silvestro, 2020

Figure 52 H^+ the only to pass from the Bubble Universe into the thin layer between the conductive spirals and the selective SO^{3-} Nafion coat (pores ⌀ = 40 Å) – Image source: http://www.fuelcellstore.com/membranes/nafion).

Doctor blade/decal transfer, hand painted, air sprayed, pulse stray swirl, ultrasonic sprayed swirl or ink jet printer **deposition** monodispersed conductive metal nanoparticles [(e.g., Cu = 5.1 nm ± 0.2 nm, Co = 4.6 nm ± 0.2 nm and Ni = 4.7 nm ± 0.2 nm visualized at Transmission Electron Microscope (TEM)] in the Nafion covering the Einstein-Dirac cones can be electro-deposited (Tin Wang et. al., 2008), towards oxygen reduction in phosphate buffer [PO_4^{3-}] = 0.1 M (pH 7.2) (T. Selvaraju, R. Ramaraj, 2005), let flow through them the spiral electricity, becoming super in the apex due to Josephson effect and promote proton (H^+) transport between Nafion PEM ionomer and catalyst.

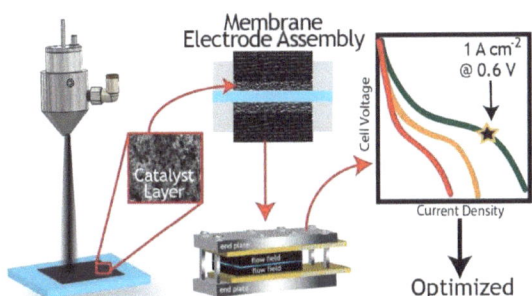

Figure 53 Catalyst deposition onto membrane for fuel cell - Image source : Tin Wang et. al., 2008.

Spin Exchange Relaxation-Free (SERF) **magnetometer lasers** could be used for detecting interaction between alkali metal atoms (Li, Na, K, Rb, Cs, Fr), secreted by aero-thermophile LUCA-descendant in the hot vapor ($H_2O_{(g)}$) at T = 180 °C expulsed form the electro-catalysts Nafion-deposited providing electronic pathways in the cones, and the Singularity magnetic field within the WormHole cylinder, preserving its azimuthal angular momentum (L) without decoherence, otherwise, caused by spin-exchange collisions of the quenching gases such as N_2 or He superfluid, that change the hyperfine stable state, shifting and/or splitting the structure of the atoms in the artificial Bubble Universe cryocooled, in which the cyclotrons orbiting toward the cathode Black Hole (-) would be polarized splitting the high complex mixture of quanta photoelectrically generated in stable racemic spooky photon spinning in equilibrium among them in the orbitals acting at a FF distance, pulsing monochromatic light, neither constant nor continuous, incident onto the Einstein-Dirac cones in accordance with the Stokes's rules, measuring the current applied (I = 100 mA/cm^2) with each pulse lasting t = 15 min at constant voltage (0.4 < ΔV < 0.9 V - Potentiostat), 50 < T_{cell} < 80 °C, 30 < T_g = 64 °C, RH = 50 % and ambient pressure in H_2 : air = 2 : 2-10 stoichiometric flow for designing a high performance PEM Fuel Cell in which the frequency changing through photo-luminescence of the LUCA descendant metagenome in the Singularity would be suitable also for its sequencing. Hence, hydrogen generation and genome sequencing via *Optical Density* (OD) spectrophotochemistry (e.g., monochromatic inelastic scattering Mira) of the microorganism using it would be paired in a unique productive process where the phonons of the hydrogen would carry the complex resonant excitation to travel FF teleporting the light microbes.

The central insulated ring enclosed before the development into conical spiral would reduce its volume Max Planck radially with the decreasing of the gaseous pressure in spherical synthetic Bubble Universe with a higher velocity than the Einstein-Dirac walls, hence, due to UHV transitioning the configuration from cylinder to bicones, which surface would present angstromscopic bubbles due to the warm moisture absorbed increasing the reflectance of the quantum of light emitted by photoelectric effect of the electrons flowing through the superconductive spiral coils electrodeposited in the transparent Nafion.

The overall apparent water clock bringing the memory of time travel back on time with its upside down turning, in an instant neither a cylinder not a double cone, but a dynamic structure changing within the Minkowski spacetime in which the evolution of LUCA let to the diversification of *Archea*, *Eubacteria* and *Eukaria*.

May have been the walls of the Einstein-Dirac bi-cone of silica-aerographene that under extremely hot plasma current have been expanded till breaking throwing all its fragments apart within the gaseous, but empty Bubble of the Universe enveloping it leaving just the Singularity and from its again, falling into a Supernova bending the Minkowski spacetime due to its infinite gravity in a new Black Hole? Well, would be the agitation of the noble, hydrogen, and helium primordial medium in expansion due to the heat generated on the walls by the cyclotrons spinning and translating capillary along its walls increasing their escape velocity $(v_e = \sqrt{\frac{2GM}{r_s}}$, where: G = Henri Cavendish gravitational constant $=\approx 6.67 \cdot 10^{-11}$ m^3 · kg^{-1} · s^{-2}, r_s = Shwarzchild radius, M = mass) moving farer form the origin decelerating, right this, the probable raising event that showed the Singularity within the clepsydra of the Einstein-Rosen bridge would have characterized the beginning of the whole astronomical evolution.

Brain works up to 11 dimensions, its complex web of connections used for thoughts and memory are yielding to mathematics of algebraic topology (Andrew Masterson, 2017). Matrixes erupted from empty mind would let weave the electricity, plasticity, potential and all the other nervous conditions shaping its vacuity as its Blood-Brain Barrier (BBB), structural and functional wall that protects the Central Nervous System (CNS) from invasion by blood-borne tiny and lipophilic pathogens releasing molecules, atoms and elementary particles which selectivity for the passage across depend on mass, charge and spin as the hypersurface of the present dividing the Einstein-Rosen bridge in two half in the space of your understanding of the folding that would have been making them one only. 38 nm UO_2 aerographene pores of the *Zeus* Singularity would carry nanoparticles, ceiling within it the descendant of LUCA elected for the teleportation across the point, a line, plane wave separated the stretched Einstein-Dirac bicones. But how long it lasts the present when you are lost in finding the clock for assessing it? Certainly, your soul will let you sense what and in which amount can overcome the threshold. Everything in one only instant? Well, during folding of the bi-half, it may be. The folded cellular plasmalemma, thylakoids, mitochondrial crest, tonoplast, ER, Golgi Apparatus, nuclear membrane are characterized by variable random electrons leaping across themselves generating potential differences calculable according to the Walther Hermann Nernst equilibrium equation, that for the neuron's membranes reach about the $\Delta V = \Delta V^0 \frac{RT}{nF} \ln \frac{[ion\ ox]_{out}}{[ion\ red]_{in}}$ 100 mV, where: R = Universal Gas Constant = 8.3145 $\frac{J}{mol \cdot K}$, n = number of electrons, F = Faraday constant

Certainly, you may understand that stay still would be the choice of the subatomic particles both just generated and the wisest that knows the astrochemical traffic jam determined to change according his vision of the right other issues would maintaining itself unmovable within the CWB, in the special relativity, old as it. So, the first and the last would be the same, one with maximum number of shells, the other naked changing configuration among them instantaneously, varying mass, charge and spin via the Einstein-Rosen bridge becoming the youngest as the twins moving from Black (-) to White Hole (+) against the chaotic mass of electrons of which the entropy and weight is sin doubt know for the infinity gravitation event like they are. Transferring subatomic

'Wormhole *in vitro*' © Copyright Antonio Silvestro, 2020

particles, atoms, molecules, compounds or microorganisms between the Einstein-Dirac cones with different probability of being in **instantaneous states** (W):

$$W = (V_c / V_S)^n$$

Where:
V_c = Einstein-Dirac cone volume
V_S = part of the volume = Singularity
n = number of movable points = teleported LUCA descendant, DNA or reference homeobox genes

By applying the *Boltzmann principle,* the energy of a chaotic system like the Bubble Universe would be equal to:

$$S-S_0 = R (n/N) \ln (V/V_0) \Rightarrow E = \beta V R (n/N)$$

Where:
R = Universals Gas Constant = 8.3 J mol^{-1} K^{-1} = 30 mJ/mol °C

The spooky particle at a distance passing through the narrow silt of the omnidirectional telescopic Singularity and White Hole, reflected by the insulating WormHole cylinder, varying the **radiation entropy adiabatically** according W. Wien (S = V $^\infty$ 0, where: V = volume), onto the Einstein-Dirac cone of the Black Hole diffracting through them easily than passing through overlapped pores would reveal according Joseph von Fraunhofer Photosynthetically Active Radiation (PAR) triggering evolution of the *Archean, Eubacteria* and *Eukarya* descendent of LUCA.

The timeline would vary breathing within the environment, when hydrogen meet oxygen in the memory of the universal solvent, which amphoteric property make it capable to dissolve the solute neutralize the medium displacing through fission, fusion and covalent bond, in the rancid past where dark matter of confusion of left behind trying to solve that meeting left in sensing the opposite way. **Water** should have been detectable in the hyperspace on the other side of the Big Bang, in the white Hole related to the Black that formed after the Supernova of the Solar System (SS), bringing life on Earth, LUCA reducing ferredoxin using hydrogen on the borders of the Singularity thriving along electrochemical potential-dynamic regulating its stability, changing with redox reactions empirically assessable in an *in vitro* simulation using Cyclic Voltammetry (CV) graphic the highest current correspond to the half inverse to its absolute value on voltage function:

$$I_{max}/(\Delta V) = -1/2 \, |I|/(\Delta V) = 10 \, \mu A(-5 \, V) = -2 \, \mu A/V.$$

Plasma is a state of matter done of an ionized gas consisting of positive ions and free electrons in proportions resulting in no overall electric charge, typically at low pressures (as in the upper atmosphere and in fluorescent lamps) or at very high temperatures (as in stars and nuclear fusion reactors). This state of matter would have been characterized the depressive and hot generation's womb, the primordial Bubble Universe about at 13.8 billion years ago when the ElectroMagnetic fundamental force originated in the Singularity would have been directionate against the walls of the White Hole attracted by non-ohmic resistant (R $\propto$ 1000 ohm) elementary particles gating the protons (H$^+$) through channels of ø = 38 nm, rectifying their AC ($\Delta V \propto$ 4 kV) into DC in plasma discharge filaments at high-frequencies (v $\propto$ 35 kHz) pulsing onto the Einstein-Dirac cone inner walls, condensing onto them (C $\propto$ 150 pF) in a thin lenticular layer, triggering the roto-translation and photoelectrical effect along them, capillary glued by weak and strong forces, descending against gravity to the accretion disk for travelling back on time as the Einstein twins. Just some Zeus monochromatic lightening would flash in the attenuating 4-He superfluid (T$_{He \; superfluid}$ = 2.17 K) in the Bubble Universe passing through the cone and the one order thinner Nafion network impacting onto the mu-metal cylinder walls and among them some pulsing discharges could even

'Wormhole *in vitro*' © Copyright Antonio Silvestro, 2020

collide on the half-cell surrounding the anode (+) travelling instantaneously from past to future reflected onto the ferromagnetic cylinder walls at straight corner electron excitation emitting spectrum-fluorometric radiation teleported into the Black Hole Singularity. So, you may understand that the overlapping of the channels among the Singularity and the Einstein-Dirac cones, as reflectance, scattering, UHV composition, Minkowski spacetime curvature are essential in determining the infinite radial pathways for the metagenome of LUCA in the Singularity photon transduced in the water.

As evolution is based on the adaptation of the tools available, the MT being a zero-waste niche novelty effective and efficient in each biome on the planet Earth, and, why not also on Mars colonized in less than six years (2025), would implement the development of the still unofficial new human species *Homo atm,* facilitate them in intra- and inter-planetary displacement making easier biotechnology deliveries. The MT, based on Microbial Electrolytic Cell (MEC), could be designed in a way that the sample in the Singularity could be teleported from past to future, and, perhaps, also back on time, fighting the progressive way of Saturn that fear the mortals commonly assessed since the analogic clock invented by *Christiaan Huygens*, stealer of the idea of 'time-telling device' felt into Galileo Galilei.

The information about the teleportation has its memory veiled in the past as in the future of our own subtle bodies related to time, unalerted in the empty space is the confine between being full or not, while, changing according the progressive present along the plane wave where the Singularity is. The plane become special and flat when the Minkowski spacetime relax itself, opening the gate for the passage of the Singularity from a hole to another, while, at the contraction would let to its undulation and ceiling of the generative spheroid in the Black Hole, as the White Hole would emit it out spontaneously. So, how would we know our primordial past? Well, just predicting the end of the Bubble Universe according the information rectified by our pineal gland.

Among the possible border lines of the continuous destiny of the Whole existing, there is mere senescence in an implosive collapse due to the overcoming of the critical density (ρ_c = 5 H_2/m^3) by the ordinary matter (0.2 < ρ_0 < 0.25 atoms/m^3), during which the temperature would asymptotically approach the absolute zero (T = 0 K) decrementing in a state of no thermodynamic **Gibbs free energy** for which dynamic corpuscles capable to displace in the empty space spontaneously with minimal effort along the Mean Free Path (MFP) where just coherent constructive interference could be detected and traced using marker gaseous particle in the medium, is equal to:

$$\Delta G = - n F E$$

Where:
n = number of electrons transferred
F = Faraday constant = 96.5 J/V mol
E = Electro Motive Force [N]

The consequent entropy increasing (Heat death of the Universe – **Big Freeze**), perhaps, related to the helium superfluid (4-He) in the Black Hole originated form the transitioning decay hydrogen from the White Hole, or the end due to over-expansion and consequent separation of the hyperboloids of the Einstein-Rosen bridge losing the Singularity in the Phantom Dark (**Big Rip**). Alternatively, if the $\rho U \leq \rho_c$, stars would burn-out be leaving white degenerate dwarf, electron-decline stellar matter core remnant and the strong gravitational mass tramps Black Holes would evaporate by emitting **Hawking-Bekenstein–Zel'dovich radiation** a positive feedback of continuous corpuscular generation among with potentially $H_{2(g)}$ exchanged form $2H^+_{(aq)}$ and $2e^-$ from the Paul Dirac Electronic *Cône – Albert Einstein Light Kegel,* freely the Singularity in the Bubble Universe. You may have

'Wormhole *in vitro*' © Copyright Antonio Silvestro, 2020

felt that just present exist and the cones are imaginary supporting the complexity of your memory, the unsustainable, unexploitable, the border between creation and destruction that characterize our *Sthula* and *Nirvana* subtle bodies which path will be stopped many times, when the threshold would be close, but not overcome as the hypersurface of the present, but life keep being immersed in the null where, being integer, 4 H_2 in 1 L of UHV of the Bubble Universe surrounding us the suitable amount for teleporting the Singularity from a WormHole to another would be done of eight electrons filling the empty hole of valence of an oxygen (O), the receiver of the memory's drops.

The quantum information encrypted in the genome of the microorganism elected for the teleportation would be polarized by the concave lenses, outward the cone in the Black Hole (-), while, inward the cone in the White Hole (+) onto the common vertex where the Singularity could pass when the plane wave of the present is stretched by the Minkowski spacetime itself varying depending on the vacuum degree of the empty space of the cylinder selectively filled by the gases protons H^+ in the Black and White Hole Dirac Cones spinning around their magnetic pole at ground state, but under the external magnetic field generated by CWB changing angular momentum, rotating in the opposite direction, passing through the Nafion and cryocooling He superfluid from the Einstein-Dirac Cones of opposite polarity, the Black exchanging $H_2(g)$ with the Bubble Universe, while, the White separated by it by a thin tension-active surface layer of mirroring H_2O ceiling the memory of the Origin of Life defined perimetrically by the accretion discs perpendicular to the relative jets, photolyase by the electrons striped-out emitting quanta by Einstein photoelectric effect from the super-conductive coils (e.g. $YBaCuO_2$) etched in the transparent cones Nafion PEM leaving in the Bubble Universe $O_2 + 4\ H^+ + 4\ e^-$, predicting aerobian organism in the hyperspace, perhaps, breathing in smaller bubbles congruent with their size and affected by sonoluminescence due to the CWB, the same that could be used for highlighting their position.

Despite the previous Albert Einstein, Boris Podolsky, Nothen Rosen – 'EPR experiment' made in the 1935 for eliminating quantum theory stoicisms from the relativity duality wave- particle characterizing light, Alain Aspects proposed in the 1980s demonstrate the bizarre separation of two photons of opposite spins that simultaneously recombine on a coincidence reflective sheet detector placed 13 meters far from the source in a phenomenon coined as **'non localité-quantic'**. This instantaneous phenomenon is in accordance with Niels Borh quantum mechanics, and shed the light on the 'spooky action' of the bosons-photons in needs for metallic laminar detector for impressing in their fulminic co-presence. Hence, despite beigns massless also photons can be managed, with a proper efficient hosting detector, as fermions (particles with half-integer spin following the 'Fermi–Dirac statistics') leptons-electrons in orbitals, regulated instead by the **'Pauli exclusion principle'** a widening concept of quantum corpuscular complementarity phenomena in the sub-atomic universe, for which the host orbitals host just to opposite sign guests at once. In brief, massive or massless of opposite charge meet themselves in a common domain as the Black (+) and White Hole (-) particles recombine in the hypersurface of the present where their overall charge is neutralized. The stretchable plane wave would be done of salty bubbles dissolved levitating in the gaseous medium under the influence of the CWB induced magnetic field (E_{CWB}) spinning the Singularity dipole between White and Black Hole, of the Einstein-Rosen bridge. Between your ass and your halo would pass an imaginary cylinder, before the stretching executed by the general Minkowski spacetime widening top and bottom with the heart respecting the present. On contrary, repulsion happen when there is confinement, for which the Singularity plasma filament are characterized by stable dual corpuscle of red-violet wavelenght, leaving the red within it and the violet out in a not combined configuration. So, you may understand that the spheroid in which LUCA where content would have physio-chemical properties similar s-atomic orbitals in which light is reflected due to photo-electric effect.

Greenberger-Horne-Zollinger state (GHZ) theorem of multi-particle entanglement, quantum cryptography with entangled photons for which complex organism could be teleported in their multiple quanta in related to satellite for Joseph von Fraunhofer Far Field over Anton Zollinger in 1998 provided the final test of Bell's inequality closing the communication loophole by using superfast random number generators an developed a method to assay dynamical diffraction and coherent superposition energy shift a De Broglie multi-wave upon neutron beam through crystals entangled reactors like the fissed n_0 from the Bubble Universe expanding the styropyrene over the Singularity in probable recombination between the Black and White Hole.

EPR qubits pair in **Bell state** ($|\Phi\rangle$), the simples' example of inequality, four specific maximally entangled quantum states combination of two qubits, which entanglement degree is measured by the von Neumann entropy of density matrix operator suitable for the chaotic styropyrene foam bending the super-conductive $YBaCuO_2$ inner cylinder in a double Einstein-Dirac cone shaping the special relativity geometry into general linear with the Minkowski spacetime network.
A positive coincidence between two qubits can be described as follow:

$$|\Phi^+\rangle = 1/\sqrt{2}\,(|0\rangle_A \otimes |0\rangle_B + |1\rangle_A \otimes |1\rangle_B) = 1/\sqrt{2}\,(|+\rangle_A \otimes |+\rangle_B + |-\rangle_A \otimes |-\rangle_B)$$

Where:
A and B = qubits $0 < A, B < 1$

Just the communication between A and B could lead a stochastic value of Bell state, but the pre-determination leading superposition depend on hidden variable, as two lover, quantum bits are chosen by the destiny of the Free Mean Path where the power of their love packeted them in the Singularity never onto present hypersurface, but instantaneously crossing it from the White to the Black Hole, past to future.

As you may now, we are in love in the most stable configuration, while, on contrary we are half of it for avoiding spinning on the contrary wave respect other bringing quanta recombining with them as bosons photons do, because in us there are Fermi-Dirac leptons electrons tot that with their mass let feel us full in an interactive domain as we all know can be reversible receiver and transmitter or transceiver under the same Minkowky spacetime. As the meeting within ourselves, so the *in vitro* Einstein-Rosen bridge with two polar loopholes with H_2 next the Black (-), while, O_2 close to the White Hole (+) outer environment, the Bubble Universe, you may visualize a timeless shared glance where signals' detection become the proof of the matter teleportation via light energy phenomena, that actually is an intrinsically condition of the whole existing.

When the information to teleport is done of multiples hyper-entanglement variables in a coherent domain as in the microbial teleportation, each with different properties, characteristics, causes, effect, traits, "matrices in a Matrix in matrices", hence, in n-dimensions is possible to distinguish $2^{n+1} - 1$ classes out of $2^{2n} = 4^n$ Bell states. Do not stop dream that every one of us could be teleported one day, but remember that the complexity of the issue depends just on how many layers have been placed around the heart or comparative anatomy. The **Bell inequality** experiments based on two others principles: locality and local realism, doesn't depends on a prearranged environmental condition, for which knowing the hidden variables the real location could be revealed, measuring both position and momentum being limited by the Heisenberg' uncertainty principle.
Electrons from a Einstein-Dirac cone to the other would collide each other, photoelectrically emitting quanta omnidirectionally onto the spherical walls of the Singularity placed in the contiguous vertexes diffusing on the hypersurface plane wave of the present when the Minkowski spacetime would be stretched by the inflation of the syropyrene foam due to the nuclear fission of tritium and deuterium into helium and consequent atom splitting due to the chain impact of the neutrons confining the gamma rays into Ni-mu ferromagnetic cylinder. Cycling cascades of nuclear

reactions underlie essential functions of all synchronic and coherent biological systems like the bright primordial Singularity where LUCA were living into proving its bio-photonic emission of variable spectral distribution picks and radiation density. LUCA would have been fed by the photos scattered into the Singularity generated by the photoelectric effect of the cyclotrons in the spiral Einstein-Dirac cones of the Einstein-Rosen bridge.

Phonon, collective excitation in a periodic, elastic arrangement of atoms or molecules in condensed matter like the Black Hole (-), are characterized by resonance $_{phonon} = \sqrt[3]{(N)}/m$ [Hz/cm], where: N = Avogadro number, exempli gratia, phonon $1H^1$ resonate at frequency ν = 3.8 GHz at temperature T = 37.8 °C. Non-linear standing wave [stationary mode frequency : $\mathbf{v_s = \frac{c}{2\pi r} \sqrt{n(n+1)}}$] biophonic fields in the Black Hole can be mathematically modelized as quadratic function $z_{n+1} = z^2$, closely related to the fractal Mandelbrot complex number set, expressed as $z_{n+1} = z^2 + c$, when $z_0 = 0$.

Lenticular H_2O hydrolyzed release H phonons in complex excitation in the White Hole (-) mirror of the far past, oscillates between ground and excited state lying at E_{exc} = 12.06 eV under the ionization threshold E_{ion} = 12.6 eV, with a UV-wavelenght Λ_{H2O} = 100 nm including electrons able to accept external energy supply from the Bubble Universe like the CWB and corpuscles dispersed and transfer it into cyclotrons vortices of lower entropy than the incoming for safely channeling and storing it in the Dirac Cone where the LUCA Singularity have been. Metagenome and Genome, precisely phosphate of DNA backbone would undergo photo-electrically effect leaving the electrons orbits proportionally with the hydrogen bonds formations, precisely, with **plasma cyclotrons frequency** of observable electromagnetic signal:

$$\nu_c = \frac{\pi q}{2 m B}$$

Where:
q = charge [e⁻]
m = mass [kg]
B = magnetic field [T]

At dawn of time, when Eos was bringing light from a White Hole to the otherside of the fringe of creation and destruction, there were Astraeus bringing radiation back into the supermassive Black Hole from which the Solar System planets generated.

The CWB wouldn't be absorbed by the Big bang Black Hole and White Hole diffused radiation into the Bubble Universe side into which we are living into.

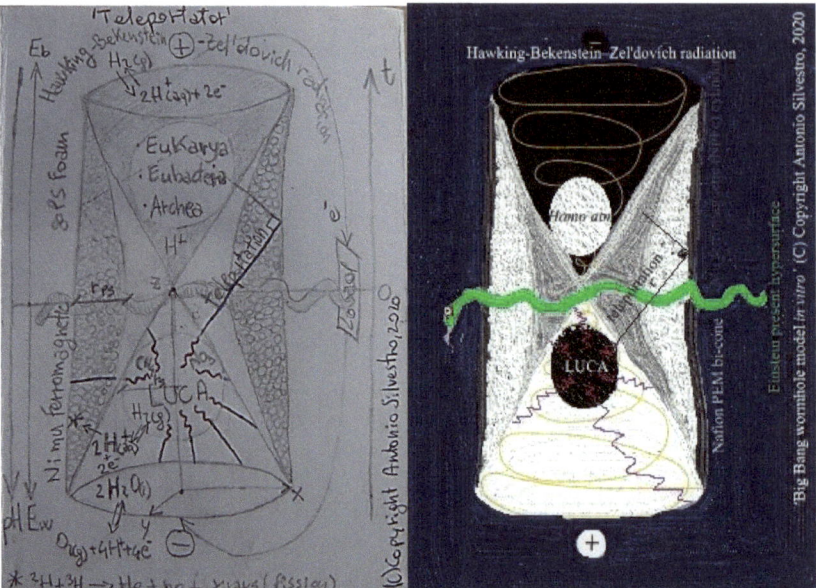

Figure 54 'Big Bang Yin-Yang wormhole model *in vitro*' graphite on paper sketch (left) and digital oil canvas (right) characterized by Mu ferromagnetic metal cylinder (Ni-Fe), Nafion PEM double cell supercapacitor shaping its biconical wall depending on the universal ESP foam inflation permitting the passage over the wave present brane of LUCA in the Singularity, filled with the organic chemical life precursor as chosen by Urey-Miller abiogenesis experiment, from the White Hole in the past before the Big Bang to the future after it, precisely, into the Black Hole of the dark Age (13.6 GYA) originated after the Supernova that gave origin to the Solar System (SS). The distance between the comoving conical electricity to the co-vertexes increase with the space expansion during the Big Bang, the electroconductive Kundalini (e.g., $YBaCuO_2$) would have its maximum at the accretion disks, while, its minimum at the origin where the positive annihilate the negative - Image source: © Copyright Antonio Silvestro, 2020.

A part Hydrogen Sulphide (HS) and Hydrogen (H_2) electron donor users *Chlorobi* before of the first *Cyanobacteria* using water photolysis for transducing light into chemical energy, they are certainly the most close to LUCA, for which using them in the DNA photo transduction experiment Wirelessly Fraunhofer Far Filed (FF) may demonstrate the presence of water (H_2O) before the Big Bang coming from the other side of the wormhole, the theoretical White Hole connected to the Black Hole form which the Solar System Supernova developed into the planets among which the Earth on which we humans, *Homo sapiens* and superior, but still unofficial *Homo atm* live on.

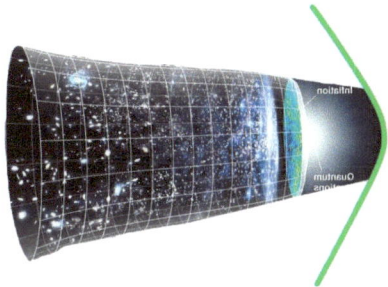 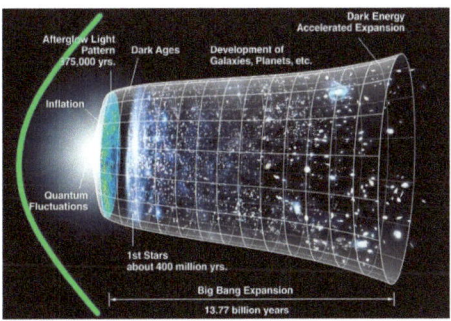

Figure 55 'Big Bang Yin-Yang wormhole' – Image source: © Copyright Antonio Silvestro, 2020, photocollage using the 'CMB Timeline300 no WMAP' by NASA/WMAP Science Team https://en.wikipedia.org/wiki/File:CMB_Timeline300_no_WMAP.jpg

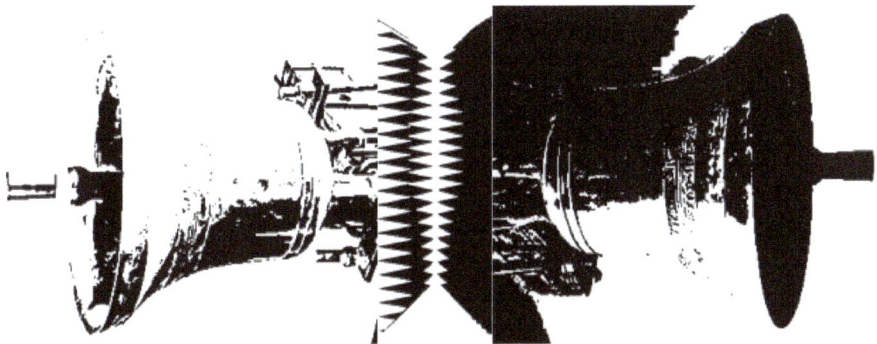

Figure 56 Two specular resonating bells, one the reverse of the other, in which the sphere is not hanged by a continuous oscillating pendulum, but levitating in the empty space free of air, defined by the infinity gravity singularities, dumb when the two antipodal spheroids would orbit around the pair of Black and White hole horizons, falling in opposite directions to a common vertex along a conical path on the inner walls of the Einstein-Dirac light-electricity bicones, annihilating their destructive frequency in the common Coherent Domain (CD) of the Bubble Universe filled by the Cosmic Wave Background (CWB), where the enantiomeric elementary particles, in the simplest quantum entanglement Bells qubit states (00, 01, 10, 11), placed in the gravitational holes are symmetric to the Einstein hypersurfaces of the present, chirality mirror plane of no sound (SPL = 0 dB) and no life as in an anechoic chamber the real shape of the plane imaged by Sir. Albert Einstein with his field (EFE) with the so described quantum fluctuation at the origin date Big Bang (13.8 GYA) – Image source: © 'Big Bang resonating bells' collage model © Copyright Antonio Silvestro, 2020 https://mmwavetest.com/modular-mmwave-antenna-test-range/

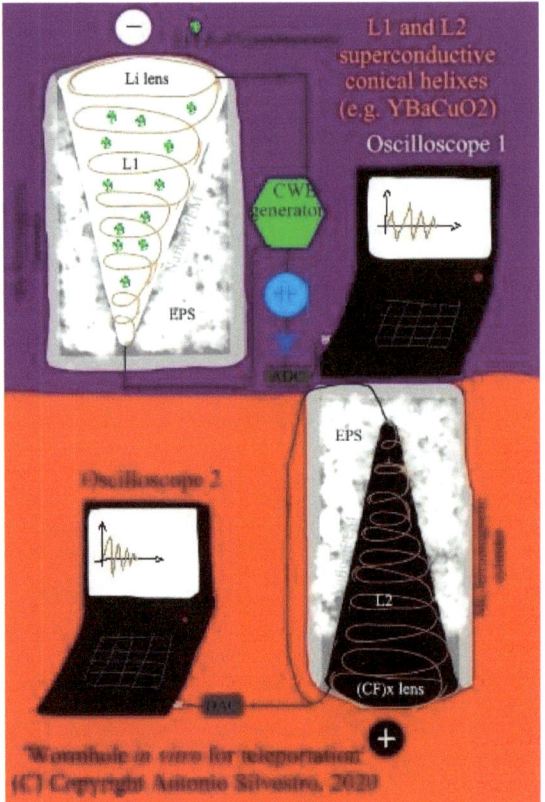

Figure 57 'Wormhole *in vitro* for microbial teleportation' N.B. For oscilloscope alignment you may find useful the algorithms in 'Hera the lovely resonator for rebirth from Sudden Circulatory Death (SCD)' (Kindle eBook 33.89 € https://www.amazon.com/dp/B08B4YBD1Q) – Image source: © Copyright Antonio Silvestro, 2020.

Teleportation Cell (TC) scale up:

1. Modified Eppendorf tubes with Nafion electrocatalyst membrane spin column in microplates for organic molecules V = 1.5 mL
2. Bioreactor for cells and microorganisms V = 125 mL
3. Bioprinter for tissues and organs V = 3 L
4. Cabin for an average human size V = 62 · 3 L

Hopefully one day would be possible to print an all-human body using -omics profile, teleportation cabins and quantum computers. **Human teleportation** could manifest itself when the organism to teleport is within the Singularity that overcome the fringe of the Einstein present hypersurface from White to Black Hole would travel in the spacetime via the 'WormHole *in vitro*'. The WormHole would act as a **transceiver** exchanging quantum information of probabilistic complex wavefunctions colliding among themselves in a coherence state from the receiving Black (+) to the transmitting White Hole (-) within itself or in series-parallel contiguous Einstein-Rosen bridges from a Universe to another. The wormhole can be seen as a high-voltage power source transformer having two inductive coils, one in the Black other in the White cone converting real, like lepton electrons, for virtual particles in the UHV Bubble Universe. The prototype Teleporter device that could be utilized for transport Fraunhofer Far Field (FF) serial, standard objects like money and coded microorganism descendant of LUCA through the Einstein-Rosen bridge *in vitro* into a fluorescent, insulated and sparking Singularity of unique designed based on the one that gave origin to the Bubble Universe where we have been spinning with our *alter ego* in the contiguous parallel Universes among which humans, both *Homo atm* and less researched *Homo sapiens* may be teleported.

Figure 58 Saiyan Son Goku (Japanese: 孫 悟空) practicing instant transmission, why humans wouldn't too?
- Image source: DragonBall Z manga by Akira Toriyama.

Human teleportation could hypothetically be made decomposing a freeze-dried body (*'Gaia lyo-grinder toilet'* by © Antonio Silvestro, 2020 https://www.amazon.com/Gaia-lyo-grinder-toilet-Antonio-Silvestro-ebook/dp/B08D6RBJ62/ref=sr_1_13?dchild=1&qid=1596732888&refinements=p_27%3AAntonio+Silvestro&s=digital-text&sr=1-13&text=Antonio+Silvestro) via sublimation let evaporate the water within the cells for later breaking them down (E = 3 GJ needed to break down all its corpuscles, ≈ 2 MJ/kg) pulverize it with the powerful *'Ares shredder'* in 'Recycled and Recycling Olympic 3D Bio-Printer (Recyclebot and RepRap-based) (Kindle eBook 4.53 €, Paperback 7.66 € https://www.amazon.com/gp/product/B08D99YWB4/ref=dbs_a_def_rwt_hsch_vapi_tkin_p2_i3) for later let it flow through the *'Teleporter'*, reassembly its elementary particles via quantum computers with metrical softwares and holographic scaffold in which the extensive traits are labeled. Potentially existing faster-than-light elementary particles (tachyons) or at least detected luxons travelling at speed of light would carry the regeneration through the void permitting the relocation on Earth, from it to the outer space and from one planet to another in the Solar System (SS) in a dreamful Teleporter.

MT 3D printed, should be designed with 2^n coupled cells in **resonance** (1 -> ∞) arranged according an oscillatory path (~), each resonating within itself and symmetrically centered as the *Sthula* and *Nirvana* divided in the middle by the *Manas Sharir*, along a bacillar, or honeycomb offset minimal free path quicker on the hypersurface of the present on the Einstein-Dirac cone shared by the chambers. If the 2D plane would be bent of β = 45°, according the Galilean physics would be faster reached the destination cells travelling with the gravity finding a stable equilibrium there. Whenever the rooms would decrease in size along the way losing shells like rockets do when moving through the envelopes of the planets, so the punctual corpuscle contained in α would reach ω. But, in the case of the MT, no hard-solid layers would manifest the inner being of the singularity, the fused Mercury (Hg), but the sonoluminescence vanishing the Bubble leaving just light in all its radiation that made LUCA live. Each resonating chambers placed onto a planetary stirrer simulating the *Seed of life* esapetals motion and each WormHole chamber pair would be inter-connected wirelessly.

Glass or bioplastic *gravimetric siphonal rack* (height a >> h) into which well would be placed the paired conical chambers Black-White Holes, each under Ultra High Vacuum (UHV) using one of the introduced vacuum pumps in *'Amphitrite and Poseidon - Vacuum pump and/or compressor'* (Kindle eBook 0.86 € https://www.amazon.com/dp/B08D8H94K1).

From Mercury 1st to Jupiter 4th, Venus 2nd to Neptune 7th, Mars 3rd to Uranus 6th, with the Saturn 5th observing and critic so the hepta-corpuscula wave of the planetary Kundalini would conduct electricity and induce magnetic field in a probabilistic oscillatory motion along the plane wave of the hypersurface of the present, perhaps, **ellipsoid** in the Minkowski spacetime. The gravitational difference in the Laplace steps when falling on the plane wave hypersurface of the present, ondulating in series is till the flexus for keeping the motion in parallel along a sinusoidal wave pathway, enhancing the tension in each step forward along the roto-traslatory displacement of the synthetic Einsten-Rosen bridge. Connect a Black Hole in resonance with a White of another Eistenin-Rosen bridge would let, perhaps, to the understanding on how rebirth people.

Figure 59 *'Einstein-Rosen bridge resonating Multi-Universe ellipsoid'* describing the parallel Universe in inter-connection on the ellipsoidal hypersurface of the present where instantaneous spinning can let the guest change according a pre-determined free will in black or white. In other words, a theoretical minimal path connecting an infinite series of Big Bang happening in continuous instants in the Minkowski spacetime endlessly onto an ellipsoid "Bubble" Multi-Universe - Image source: © Copyright Antonio Silvestro, 2020.

 Dr. **Antonio Silvestro** Ba in *'Biological Sciences'* got with honours and joiner of the MSc *'Plant, Food Sciences and Environmental Biotechnology'*, divulge his holistic wisdom as Kindle Self Publisher (KDP) on Amazon. Erasmus in Spain and Erasmus + in Latvia, Europe, Asia and Africa traveller (33 countries visited), took part in permaculture, agroforestry, aquaponics, and phytoremediation eco-projects in countryside with developing communities,

in organic farms, designing environmental and wellness architectures, awaking souls via Yoga, Tantra and Shamanism spiritual healing retreats, for example, giving the *'HolYoga'* workshops in Denmark and Germany, both in private estates and in public festivals. He is the designer of the *'Telepika – Telescopic Pipe Moka'*, discoverer of the 'Tezcatlipoca Black Hole' determining the transition Triassic – Jurassic, named by as the Aztec God that guided himself in revealing it behind the glyphs of the 'Sun Stone', theoretical modeller of the Ab Initio Molecular Dynamics of the Solar System (SS-AIMD) and of the Big Bang based on a Wormhole *in vitro* invented by himself.

www.ingramcontent.com/pod-product-compliance
Lightning Source LLC
Chambersburg PA
CBHW040220220526
45473CB00001B/65